D-Brane

Koji Hashimoto

D-Brane

Superstrings and New Perspective of Our World

Dr. Koji Hashimoto
RIKEN
Mathematical Physics Lab.
Hirosawa 2-1
351-0198 Saitama
Wako
Japan
koji@riken.jp

The Original Japanese version was published by University of Tokyo Press, Japan, 2006.
Translated by Haruko Hashimoto and Koji Hashimoto.

ISBN 978-3-642-43181-4   ISBN 978-3-642-23574-0 (eBook)
DOI 10.1007/978-3-642-23574-0
Springer Heidelberg Dordrecht London New York

Printed on acid-free paper

Springer is part of Springer Science+Business Media (www.springer.com)

To Haruko and to my parents

Preface

This book is dedicated for explaining and introducing "D-branes," a new concept on which theoretical developments in superstring theory in the past decade are all based on, with brief words and equations. The original version of this book was written in Japanese, and published by Tokyo University Press in 2006. I hope you enjoy surprising and intriguing novel perspective of our world which is shown by frontier of researches on elementary particle physics, cosmology and superstring theory which may unify all matters and forces of our world.

D-branes, or in general "branes," are membranes in higher-dimensional spaces, in mathematical physics. Once we apply this notion of "branes" to various physics, such as elementary particle physics and cosmology, we reach surprising change of our perspective on our world: we may live on 3-dimensional membranes in higher-dimensional space. This scenario is called "braneworld," and not a scientific fantasy. In fact, every day and night, quite a number of particle physicists and cosmologists all over the world investigate this fascinating possibility of "living on hypersurfaces." Why is this fascinating? It is not because the hypothesis is novel as a sci-fi, but because it is a logical consequence of superstring theory, a candidate theory unifying all forces and matters. And the hypothesis has a scientific possibility of resolving long-standing problems in particle physics, cosmology and gravity.

This book is written for undergraduates who are interested in particle physics, superstring theory and cosmology, and also for researchers and graduate students whose major is not particle physics. In writing this book, I tried to minimize the assumed knowledge for readers, and I skip subjects which are necessary to really study elementary particle physics. I suppose readers have basic knowledge only on mechanics and electromagnetism, and a short path to the frontier research results is provided in this book. As a whole, I describe, in plain language, mechanism of elementary particle physics and superstring theory, and bring a "flavor" of frontier science research on D-branes. For instance, many conceptual figures are used in this book, for you to grasp easily and intuitively various physical concepts. Indeed, researchers everyday draw these kind of figures on blackboards for research.

Footnotes are for advanced/supporting explanations so that most of readers can just skip them on a first reading. Some of the important concepts which are indispensable in undergraduate/graduate courses in physics, such as quantum mechanics and quantum field theories, statistical mechanics and thermodynamics, are not described in detail but briefly outlined in this book, so readers who like to learn particle physics on more solid basis are advised to use some other textbooks. But I hope that this book is an entrance to view easily what is actually happening now at the frontier of theoretical physics.

This book is not a textbook, so readers can just skip whatever they feel difficult or not interesting. I had to use some equations to show the fun of logic which leads to tremendously interesting results of the physics researches and how researchers are excited with them. But equations are only for advanced readers, so it is not necessary to follow all of them. It is important to "feel logically" that D-branes are necessary and indispensable in superstring theory, and how they provide us with such an interesting outcome. Even if in some parts of the book you may feel that the content is just mathematics, it is important to notice that all physics are based on mathematical equations, and the "brane paradigm" is completely based on such a logical reasoning.

D-branes are a kind of "solitons." Solitons are one of the important concepts in physics, so the first part of this book is for explaining basics of solitons. You will find that the importance of the D-branes originates in solitons. In case you are bored by the mathematical structure of solitons, you may start reading from Chap. 6. Braneworlds, a hypothesis that we live on some hyper-surfaces in higher-dimensional space, and other applications of D-branes are treated topic-wise in Chap. 6. The topics covered there are

- Braneworld scenario
- Black hole production in experiments
- Braneworld cosmology and inflation
- Counting of states on black holes by superstring theory
- Holographic principle: gravitational description of quarks (also known as AdS/CFT correspondence)

and readers can start from anywhere. These are just a part of the main achievements of superstring theory using D-branes; there are other important developments which could not be covered in this book. Some explanations in this book are not precise in theoretical sense, which is for readers to understand physics and mathematics in an easier manner. At the end of this book, I include some web links to research papers, with which readers can just feel more flavors of frontier researches.

In 1995, I entered a graduate course in Kyoto University and eventually started reading the famous original research paper by J. Polchinski on D-branes, written in 1995. This paper initiated the revolution of superstring theory. Now, after a decade, this book is complete, largely owing to my friends and collaborators with whom I really enjoyed physics discussions, at Kyoto University, University of Tokyo, and all over the world. I would like to thank them all. I thank my parents who allowed me to pursue my interest in physics. And I am indebted to my wife Haruko for encouraging me and supporting my life as a physicist.

Tokyo   *Koji Hashimoto*
August 2011

Contents

Chapter 1
From Elementary Particles to D-Branes and Strings

D-branes are higher-dimensional membranes in string theory. The simple membranes take on special importance in physics. The D-branes are not only widely applied for various subjects of physics related to string theory, such as elementary particle physics, cosmology and mathematics, but also might be fundamental constituent elements of the whole universe. Looking at recent results of researches of the D-branes, we might be able to say "All matters and interactions which our bodies consist of are made up of D-branes," or we might be able to say "we live in D-branes." These two expressions apparently look quite different from each other, but they are just different ways of viewing the same D-branes. And these two possibilities about D-branes are parts of the subjects which a large number of elementary particle physicists concentrate on currently. The main purpose of this book is to tell attractive aspects of this D-brane to readers intelligibly, and to enjoy together excitement in the leading-edge study of the candidacy of ultimate theory, namely, string theory which unites whole elementary particles and forces.

The definition of the D-branes is, in a word, "surfaces (objects) to which end points of fundamental strings can be attached." The "D" in D-branes comes from the initial of Dirichlet boundary condition (fixed boundary condition), and it means that the end points of string are attached at particular locations in a space. This fixed plane is the D-brane. On the other hand, "brane" means a part of the word "mem-brane." Speaking of membranes, we easily imagine objects extending in two dimensions in space, such as balloons and bubbles. But the D-branes are objects generally extended in any dimension as well as two dimensions. For example, the dimensions span from low to high, such as zero or six dimensions. You can see this image in Fig. 1.1.

It was 1995 when the importance of the D-branes was recognized in string theory. It was found that D-branes can be fundamental constituent elements replacing the role of strings. Since more than a decade ago, string theory, elementary particle physics and cosmology around it have enormously developed under the influence of the D-branes. In particular, in these 15 years in string theory, string theorists have devoted themselves to the study of the D-branes. The reason for this is that D-branes have splendid fascination and flexible applicability in them.

K. Hashimoto, *D-Brane*, DOI 10.1007/978-3-642-23574-0_1,
© Springer-Verlag Berlin Heidelberg 2012

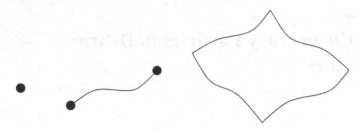

Fig. 1.1 Objects (branes) of various dimensions in space. *Left*: particles are a kind of branes. *Middle*: string-like one-dimensional brane. *Right*: two-dimensional branes are ordinary membranes

In this book, I will briefly explain string theory in which the D-branes originate, and further give a full account of what D-branes are. And then I will introduce several applications of the D-branes and their vast achievements in these 15 years, in a topic-by-topic manner. These practical researches are, of course, currently developing, and they reached the level of forming independent established research subjects, owing to achievements by a huge amount of efforts made by researchers. I would like to show how wonderful the D-branes are, through the explanations of them in Chap. 6. In the final Chap. 7, we will look at a possibility that the D-brane can be a fundamental constituent element constructing our whole world. I will introduce you to various recent challenges by elementary particle physicists in pursuit of ultimate unified theory.

1.1 String Theory and Problems in Elementary Particle Physics

String theory is a candidate of the unified theory which describes the whole elementary particles existing in this world and the forces interacting between those particles in a unified manner. To find the ultimate theory is the final goal of elementary particle physics and also is a dream of elementary particle physicists. String theory still remains incomplete, and D-branes are considered to play an important part in shaping string theory's future. That is the reason why we study D-branes. In this chapter, I will give an overview of elementary particle physics and string theory, followed by a brief explanation of what D-branes are and what is the importance of the D-branes, as an introduction to the whole contents of this book.

Elementary particle physics is a subject to describe irreducible constituents of matter, "elementary particles," which compose matter and force. Strictly speaking, the most fundamental elements are still unknown, so the elementary particle physics is also called as "particle models." However, elementary particle physics has a basic theory which explains fairly accurately results of elementary particle experiments, and it is called "Standard Model of elementary particles." This Standard Model dominantly stays in elementary particle physics over two decades. There barely exists a few results of experiments which are inconsistent with its predictions. After all, we can say that elementary particle physics is based on the Standard Model.

The Standard Model is written as a "quantum field theory" in mathematical terms. The word "quantum" means quantum mechanics, and "field" of the field theory is, for instance you can imagine, electric and magnetic fields. Each field appearing in the Standard Model corresponds to each kind of elementary particles. For example, there is a field of the electron, and a field of the elementary particle mediating the electro-magnetic interaction (it is in fact the light), and so on. In the Standard Model, interactions between various elementary particles (describing forces interacting between them and how they change their species) are included as theoretical constants (parameters). You can theoretically calculate interactions of any matters by using the Standard Model. For instance, you can calculate what kind of a final state two nuclei go to when they collide with high speed and react/scatter.

If we can precisely predict all experimental results, then what do we want further? Some readers might think that we are just fine with the Standard Model being the ultimate theory. However, the truth is that the Standard Model has the following problems. The Standard Model has no problem as a "mathematical" theory, as it gives a start point and calculation methods for describing scattering of elementary particles. However, "physically" it has problems.[1]

The biggest problem in the standard model is that it does not include gravity. In fact, it is very difficult to unify Einstein's general relativity describing gravity and quantum mechanics on which the standard model is based. In a naive unification of the general relativity and the quantum mechanics, physical quantity (for example how the gravity acting among particles changes by quantum effects) diverges, and it does not make sense at all.

Furthermore, as the second difficulty of the standard model, there are too many unconstrained arbitrary constants (parameters) in the theory. For instance, the standard model of elementary particles cannot answer the question such as why the mass of electron equals actually measured values. This is because the part relevant to the electron mass in the standard model is just an arbitrary constant. Moreover, the standard model assumes the species of the elementary particles from the first place, so it does not, of course, explain why those particles show up. It can never explain why there are two more species of particles resembling electrons, and why we have electromagnetism, weak force and strong force, as interactions.

By the way, the main theme of this book, namely string theory, is a theory regarding strings extending 1-dimensionally in space as fundamental constituent elements. The reason why string theory is a candidate for an ultimate theory unifying the whole matters and forces is that it has a possibility to solve the difficulty and the

---

[1] To be precise, the Standard Model has some mathematical problems, such as technical difficulties in practical calculations by using the Standard Model, and no proof for quark confinement which will be described later in Chap. 6.4. However, we can say there are no mathematical problem in the sense that the starting point is given in the theory. As described below, it remains a physical problem why we choose the starting point like that.

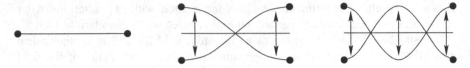

Fig. 1.2 Oscillation of an open string. From *left* to *right*, it shows a state with node numbers 0, 1 and 2. The ends of the string are subject to a free boundary condition. We can easily imagine strings with more nodes

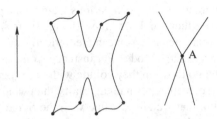

Fig. 1.3 *Left*: an interaction (scattering process) among two open strings in spacetime (the time direction is specified by an *arrow* on the *left*). *Right*: corresponding interaction for particles. At the central point A, the interaction occurs, where one has a freedom to put a constant showing the strength of the interaction

problems in the standard model mentioned above. As for the first problem, that is, the unification of gravity and quantum effects, string theory contains gravity naturally as you will see, and furthermore it does not give divergences in calculations of the quantum effects. Moreover, in string theory, just the fact that everything is made of strings determines almost all the structure, by a mathematical consistency. Hence, we have very little freedom for the arbitrariness of interactions. Finally, it is expected that in string theory there is no arbitrary constant which can vary. Therefore, once string theory is complete, it can possibly explain the mass and the charge of electrons theoretically, in the end. This offers an approach to the second problem mentioned above (Fig. 1.2).

These are the important reasons why we consider string theory. Then, how does the one-dimensional string describe point-like elementary particles? Strings extend in one dimension in space, which gives itself a degree of freedom to change its shape, compared to particles. Let us take a look at an open string. The shape of the string can be classified by the number of oscillation nodes. We can regard each of this as a particle. That is just single string can describe infinitely many kinds of particles at once. In the case of a closed string, one of the oscillation modes stands for a graviton describing interactions of gravity. This is the reason why string theory contains gravity.[2]

Let us consider how interactions of particles are described in string theory. See Fig. 1.3. This figure shows a spacetime (=space + time). The vertical direction

---

[2]This fact was found by T. Yoneya and by J. Scherk and J. Schwarz in 1974.

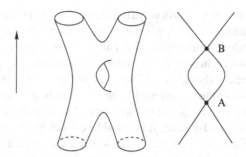

Fig. 1.4 Two closed strings propagate in space and interact. The worldsheet has a hole (*Left*), while its particle version has a single particle loop (*Right*). In calculations of scattering processes of the particles, the limit of bringing the interaction point A closer to the point B generates a divergence

shows the direction of time, and a horizontal slice stands for the shape of a string at a certain time. This membrane in spacetime is called a "worldsheet" and the motion of a string is described by this. Here, a worldsheet of a string scattering process corresponding to a particle scattering (Fig. 1.3 right) is shown, where we can immediately understand that the spacetime point of the particle interaction (the point A in Fig. 1.3 right) does not exists for the case of string theory. This means that, though it is possible in particle theory to put different parameters at the interaction points for each kind of point-like particles, string theory does not have the freedom. Interaction strengths of the theory are necessarily unified. This is contrast to the standard model which has a large arbitrariness with parameters put for each interaction, and so it is a first step to solve the problems of the standard model.

Now, let us take the case of a scattering of closed strings, for instance (Fig. 1.4). The left side shows a scattering of strings, and the right side is a corresponding scattering of particles. In the figure of the particles, if the loop at the center shrinks, the calculation of the scattering includes a divergence.[3] This is because the two interaction points A and B come closer. On the other hand, in string theory this infinity does not show up. Strings are extended, so there is no point corresponded to the points A and B in Fig. 1.4 left.[4]

In this manner, string theory provides us with a possibility to solve the problems of the standard model of elementary particles. It is wonderful that quite a simple procedure of replacing particles with strings may solve the difficulties. Then, why does string theory still remain as just a candidate of ultimate theory? In fact, string

---

[3]In particle theories, a procedure called "renormalization" exists which can eliminate these divergences. However, in the theory of gravity, the standard renormalization is impossible.

[4]This is a heuristic argument. In reality, if we consider the case of the central hole in Fig. 1.4 left shrinking, or the case of decomposing the string into infinite number of particles as mentioned earlier, possible infinities may show up. However, string theory has an open-closed duality which we will see later, and due to such mechanism which is intrinsic to string worldsheets, the calculation results of these scattering process become finite.

theory has a crucial unsolved question. String theory does not have any starting point yet, as opposed to quantum field theories on which the standard model is based. For instance, in string theory, rules which give calculations of scattering processes for weakly interacting strings are generally known, but for strong interactions, nobody knows how to calculate them. In addition, as string theory contains gravity, it should be able to answer the question why the universe is as what we observe now. However, what we know now is just calculations of scattering of strings moving in a spacetime which we put by hand. It is considered that in string theory these two problems are deeply related with each other – it reaches a problem of how string theory is defined at the end. We know only a part of string theory.

D-branes appearing in string theory are the most important clue to solve this ultimate problem. D-branes might be fundamental constituent elements defining string theory.

1.2   What is D-Brane?

Then, what is a D-brane? In the case of the above-mentioned example of oscillation of an open string, we considered a free motion of a string by imposing free boundary condition on the endpoints of the string. One can also set a fixed boundary condition instead of the free boundary condition. To impose a fixed boundary condition equals to demand that there is a high-dimensional membrane in a space and end points of the string should be attached on the membrane. You can choose the dimension of the "membrane" arbitrarily. This is a D-brane. "D" means a fixed boundary condition: it is named after the initial of Dirichlet boundary conditions. For instance, if one considers a point-like D-brane, the string must have its edge attached on the point. Furthermore, if one considers a membrane-like D-brane of two dimensions, the open string can move only on the membrane limitedly (Fig. 1.5). On the other hand, in the

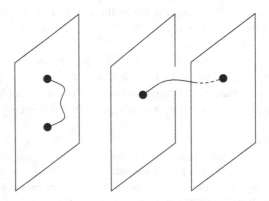

Fig. 1.5 D-branes in a space, and open strings whose end points are on them. End points of the string may be on the same D-brane, or may be on different D-branes. This example shows D-branes put parallel to each other

case of closed strings, they can move in the whole space freely because they don't have endpoints. That is, motion of open strings is confined to the "limited space" which is the D-brane.

By applying this concept, we come up with an interesting idea called braneworld. For example, let's imagine a situation where the whole space is not 3-dimensional but higher-dimensional, and the dimension of a D-brane in it is spatially 3-dimensional. And let us assume that matter constituting our body is made of open strings. Then, it is concluded that we are confined inside this D-brane, and only closed strings, namely, only gravity can propagate in a higher-dimensional space! This scenario is called "braneworld," following the sense that our world turns out to be a D-brane itself. The concept of the braneworld is used not only for D-branes in string theory but also generally when extended objects are introduced in higher-dimensional space. The multi-dimensional extended objects are called just simply "brane" without "D."

Let us suppose that D-branes of various dimensions exist at various directions in the whole spacetime. Then, as there are open strings connecting each D-brane, we can label the open strings. For instance, if two D-branes cross each other, open strings connecting them must be localized around the intersection points because of the string tension. Different kinds of particles come out from the oscillations of various open strings. In this way, by distributing various D-branes in a high-dimensional space geometrically and well-organizedly, one can rebuild the standard model of particles. This method gives us geometrical and interesting possibilities to explain the standard model of the elementary particles by using the string theory (Fig. 1.6).

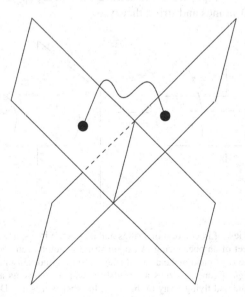

Fig. 1.6 D-branes crossing in a space. 2-dimensional D-branes share a line and also cross each other. An open string connecting two D-branes localizes around this line

On the other hand, D-branes themselves are dynamical objects. For example, D-branes have masses and they can move with some velocities. And D-branes influence each other, as they scatter, collide or create a bound state after combining.

In the view of a brane world that the universe of our 3-dimensional space is the D-brane, what kind of effect does the motion of a brane in a high-dimensional space provide? In fact, this turns out to reproduce a time evolution of the universe – how the universe expands as time passes. Furthermore, there is also an interesting research in which a big bang which is the origin of the universe is regarded as a collision of two D-branes.

Moreover, D-branes can be regarded as black holes. As mentioned above, string theory contains gravity, the most mysterious object made by the gravity is the one called a black hole. Black holes are remnants of stars shrinking by the gravity itself after undergoing supernova explosions. They are "holes in space" which continues sucking all matters and even the light. If one comes closer to the black hole, one feels enormous gravitational force. We can regard this black hole as a D-brane. Let us see Fig. 1.7 standing for a worldsheet of a string. The object on which an open string has its end point is a D-brane, and let us consider this open string propagating in time to form a loop trajectory. If one exchanges the time and the space directions of this sheet, the sheet shows a configuration of a closed string emitted from the D-brane and flying away to the right. Therefore, D-branes originally defined as membrane on which the end points of strings are attached, are the origins of the gravity. Because the gravity emitted spreads all over from the location of the D-brane in a spherically symmetric manner, it will get weaker as it goes away from the D-brane. On the other hand, it is very strong near the D-brane. Namely, D-branes are black holes. By using this equivalence, properties of black holes, which are mysterious objects, can be revealed by the D-branes and string theory.

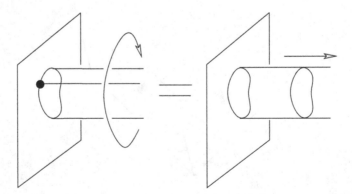

Fig. 1.7 A schematic view of interaction of strings and a D-brane in a spacetime. In the *left* hand, we consider a worldsheet of an open string. An open string which has an end point on a D-brane extends infinitely (to the *right*) and make a round (the *loop*) longitudinally in time. Regarding the worldsheet as in the *right* figure, it shows a worldsheet which represents a closed string being emitted from the D-brane and flying away to the right. In other words, the D-brane is a source of closed strings

In this way, introducing D-branes in string theory has changed our perspective on our "world" drastically in elementary particle physics and cosmology. A large amount of various physics phenomena are interpreted as geometrical distributions, motions, and interactions of D-branes floating in a high-dimensional space. This doesn't only add some new interpretations of physics established earlier. This is because, in developments in physics, new constructions and new concepts create fundamental reformulations.

For instance, there is a very interesting idea for which this D-brane gave a concrete example. It is called holography: physics in different spatial dimensions are equivalent to each other in fact. From a common sense point of view, if dimensions of spaces we consider are different, physics must be completely different. So, the holography is a quite odd and interesting situation. The very important aspect for identifying D-branes with black holes, as described earlier, is the duality. Even though we thought that we described them with an open string, in fact we could alternatively see closed strings emitting. This is a property called an open-closed duality. Now, the gravity, namely the closed strings, can propagate outside the D-branes, while the open strings remain confined on the D-branes, so, in the physics described by each of them, the dimensions considered must be different. As those different dimensions give the same physics, this is an example of the holography (Fig. 1.8).

In string theory in which the D-branes are introduced, it has been gradually understood that various "dualities" exist, related to the D-branes. The duality means a property that two physical systems are equivalent to each other, as the holography mentioned above. The most interesting duality conjectured in string theory is a duality of exchange of D-branes and strings. In various physical models, there appears a duality of exchange of fundamental particles and "solitons," and it is applied to string theory.

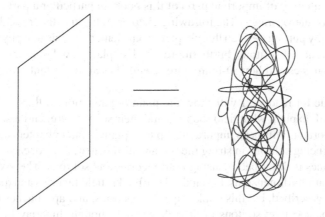

Fig. 1.8   D-branes (*Left*) can be considered as a gathering of strings (*Right*)

Soliton is a state in which a large number of fundamental constituent elements of a theory condense and move collectively as if they are a single object.

For instance, a theory having a duality of the exchange of particles and solitons is (a slight generalization of) electromagnetism. In electromagnetism, electrons have electric charges and produce an electric field around them. On the other hand, equations of the electromagnetic field (Maxwell equations) have the duality under which the exchange of an electric field and a magnetic field keeps the equations invariant. From this fact, we might expect the existence of a matter having magnetic charges, in correspondent with the electrons having electric charges. This object is called (magnetic) monopole, and is considered as a soliton constituting of a collective motion of the electromagnetic fields. That is, there is a duality exchanging particles and solitons, in this example of generalized electromagnetism.

Matters are made of strings in string theory, and a black hole is formed once a large amount of matters condense. Then, we can say that a large number of strings condense to become D-branes. That means "D-brane is a soliton in string theory." If strings can be exchanged with D-branes in string theory, D-brane can be considered as a very fundamental matter. From this observation, there appeared an idea that all matters and interactions constitute of D-branes. Also it was then understood that, in this exchange, the strength of interaction in string theory is mapped to its inverse. Namely, it became clear that the situation of weak interaction we can easily understand is in fact equivalent to the situation of strong interaction. This is a big step to resolve the issues in string theory which are described above. D-branes may be fundamental constituent elements in final and ultimate unified theory describing our world.

1.3   Organization of This Book

Above is a summary of important parts of this book, in particular a part concerning the essence of the D-branes. The following chapters introduce theoretically physics of D-branes, by putting flesh on this chapter, from elementary particle physics, string theory and basics of solitons. Furthermore, I will explain how D-branes are applied for various physics and how D-branes are regarded as an important object in string theory.

As a guide for readers, I will show the main organization of this book. First in Chap. 2, I will explain what field theories and their solitons are, and also why solitons are important. This is an important step to explain in later chapters the meaning of D-branes being solitons in string theory, and also is necessary because dynamics of the D-branes is described by using field theories and solitons. There will appear solitons of various dimensions. In standard textbooks, field theories and quantization of fields are described, but this book skips those stories, and approaches them from a different perspective: solitons in field theories. Although this may look strange, I hope readers to enjoy this different viewpoint on quantum field theories. For readers who are familiar with quantum field theories, this "alternative" viewpoint may serve as a new way of looking at quantum field theories, as there are few textbooks which discuss solitons and their importance in quantum field theories.

Next, in Chap. 3, let us see a surprising application "brane world," by looking at solitons from the dimensional viewpoint. Though this needs a high-dimensional space, the high-dimension is also a concept naturally appearing in string theory. We learn the basics of string theory in the chapter and get how high-dimensional spaces are dealt with in string theory. I hope readers may enjoy how dimensions are treated in physics. In Chap. 4, let us see that the high-dimensional space and the solitons appeared in Chap. 3 become a basis of a new concept "D-branes." There a definition of the D-branes is given, and it turns out that the D-branes are in fact solitons of string theory. The aim of the chapter is to know what D-branes are, and to know the importance of them.

In Chap. 5, I will explain various aspects of dynamics of D-branes. There, it is described that how D-branes move, combine, create and annihilate. You can see vivid appearance of the D-branes living in a high-dimensional spacetime. What is more, you can also learn various interesting relation between D-branes and soliton. You need these knowledges in order to learn broad applications of the D-branes in later chapters.

And in Chap. 6, I will pick up and explain four topics as various applications of D-branes. (You can read each section in Chap. 6 independently.) The explanations there will lead readers to a mainstream of recent explosive developments in string theory. D-branes, which brought a new notion of high-dimensional membrane in high-dimensional spacetime, flourish not only within string theory but also in various physics close to string theory. In the subjects such as cosmology, relativity, black holes and of course elementary particle physics, I will give full explanations of the new methods and paradigms given by D-branes. In Sect. 6.1, I will explain "brane world," developments in higher-dimensional particle theories, and "creation of artificial black holes" which those theories predict in experiments. Then, it will be described how D-branes explain the inflationary universe and how D-branes clear up mysterious thermodynamic properties of black holes, in Sect. 6.2 and Sect. 6.3, respectively. Furthermore, I will give a detailed description of the "holography" which was obtained by the D-branes entering the story, in Sect. 6.4. It realizes a marvelous physical dualities: the holography is an equivalence between theories with different spatial dimensions, and we can calculate physics of quarks by using a gravity theory.

Finally, in Chap. 7, I will briefly take a general view of, how the string theory, which is regarded as an ultimate theory unifying all elementary particles and interactions, has approached its final goal, by active researches of D-branes. Thanks to the D-branes, the ultimate dream of physicists may come true.

I hope that through this book you may feel flexible and visual charm of the D-branes, possibility of the D-branes, and also the excitement in researches of the D-branes.

Chapter 2
Solitons and Elementary Particle Physics

As I mentioned in the preceding chapter, the D-branes play an important role in elementary particle physics and cosmology. A purpose of this book is to introduce how prominent the D-brane physics is, but another very important purpose is to make clear why the D-branes physics is prominent and why they are considered to be very important.

What is the meaning of studying the D-branes? The answer resides in the fact that "D-branes are solitons in string theory." Solitons are objects which play a very important role in field theories on which elementary particle physics is based. In this chapter, I will have an account of what field theories are, what are the objects called solitons which appear there, and what in elementary particle physics is the meaning of clarifying the properties of the solitons. In the next chapter, I will explain the basics of string theory. When we combine the solitons described in this chapter and string theory given in the next chapter, we can see the importance of D-branes, namely, solitons in string theory, introduced in Chap. 4.

The present elementary particle physics is completely based on a theoretical scheme called "quantum field theory." Standard textbooks of elementary particle physics start with explanation of how elementary particles are described by fields in field theories, and proceed to theoretical description of scattering of elementary particles by the computations of quantum field theories. Solitons are not emphasized there, because the true power of elementary particle physics resides in this precise identification of particles and their physical processes with computations in a distinguished quantum field theory, called "standard model of elementary particles." In this book, alternatively, we start by accepting the fact that there is a well-established quantum field theory called standard model, and will explore what is the problem of the standard model. Almost any field theories have nontrivial classical solutions called solitons, and so solitons are quite generic concept, though not emphasized in most of textbooks. Solitons are not elementary particles, so not

K. Hashimoto, *D-Brane*, DOI 10.1007/978-3-642-23574-0_2,
© Springer-Verlag Berlin Heidelberg 2012

directly related to elementary particle physics. However, they play important roles in resolving the problems of various field theories. I will emphasize this viewpoint in this chapter, and will see how solitons are relevant in the problems of the standard model of elementary particles.

First, in Sect. 2.1, we will have a brief look at what are the quantum field theories, and how they are related to elementary particles. There, I provide a list of problems of the present particle physics: (1) *quantum gravity*, (2) *arbitrariness problem*, (3) *hierarchy problem*, (4) *quark confinement problem*. The theme of this book is to provide what researchers are trying now, to resolve these problems, via solitons, D-branes and superstring theory. For each subject in the list, a short explanation is included, to bring you the information where you should read in this book concerning the question you are interested in. Whenever you feel that the content of the book is too mathematical, you can come back to these problems in the list in order to remind yourself of how the mathematical tools explained in this book can be used for resolving important physical questions in elementary particle physics.

Then in Sect. 2.2, the notion of solitons are introduced mathematically, through the field theory language. Solitons appearing in elementary particle physics are packets of energy moving as if they are particles (though they are not particles). Generally speaking, solitons appear in various physics, not only in the elementary particle physics, and also they are seen as physical phenomena around us. For instance, tsunami and vortices are also solitons. Precisely speaking, solitons are objects defined in theoretical physics called field theories.

In Sect. 2.3, let's follow concrete examples of simple solitons by using figures and equations, in order to help you image what the solitons are. There, it is shown how solitons appear in physics. It will be made clear that the situation of appearance of solitons correlates closely to what kinds of symmetry the physics system we consider has and how the symmetry is broken.

In Sect. 2.4, I will explain how we can actually observe and find the solitons in elementary particle physics, and what kinds of knowledge the finding of solitons brings to the elementary particle physics. In fact, the usefulness of solitons is considerably wide and not restricted to the elementary particle physics. I will explain how useful solitons are in various theories of physics, there. The new symmetry "duality" which will be made clear in that section is the origin of importance of D-branes in string theory. Among various kinds of dualities, the important duality is a strange symmetry with which a physical theory of concern is actually equivalent even if particles and solitons are exchanged. That is, if solitons of string theory are the D-branes, you might exchange strings for the D-branes, and you could regard the D-branes as fundamental constituent elements in string theory! (This will be shown in Chaps. 4 and 7.) An aim of this chapter is to introduce a way of thinking of dualities which is a very important idea based on string theory, elementary particle physics, and field theories.

2.1 Elementary Particle Physics and Field Theory

2.1.1 Fields

Field theories are used as a fundamental theory in physics, especially in elementary particle physics, nuclear physics, solid-state physics, and cosmology. Elementary particle physics among those describes elementary particles existing in our world and their interactions (how they influence each other through forces, how they scatter, and so on). It is closely related with string theory. In order to learn how solitons work there, first we shall see the notion of "fields" and field theories by using a simple example of water surface, and then will see how the field theories are used in elementary particle physics.

To grasp the notion of the "field," one of the best ways is to imagine a mathematics describing water surfaces. Let us imagine a water pond. The surface of the pond usually oscillates and fluctuates, by winds and some other external inputs. In ideal situation, we have no external disturbance, and we imagine that the water surface has a wave expanding and traveling.

What is a tool describing the height of the water surface? If each point of the planer water surface is described by a two-dimensional coordinate system (x, y), the height h of the water surface is a function of (x, y). In addition, since this height changes in time, it is also a function of time t. So, in total, the height is written as $h(t, x, y)$. Functions of spacetime coordinates like this are called "fields."

If you want to describe a motion of a single particle, then you need to specify the location of the particle at given time, so the motion is described by $X^i(t)$ where $i = 1, 2, 3$ are the indices for the spatial directions. This is just a function of time t. But, in general, fields are functions also of spatial coordinates, as in the case of $h(t, x, y)$.

A field theory called "standard model of elementary particles," which has been established to describe elementary particles in quite a precise manner, introduces a field respectively for each particle existing in the world, such as electrons and quarks. For instance, we introduce an electron filed to describe the electron. Namely, the type of the field corresponds to the type of the particle itself.

One of the most familiar field theory is the theory of electromagnetism, and this is a part of the standard model of elementary particles. Electric and magnetic forces are one of the fundamental forces in nature, and elementary particles carrying electric charges are subject to these forces. The electromagnetic forces are described by fields; these are electric and magnetic fields. And, in fact, in the standard model, there is an elementary particle corresponding to these fields: it is the photon.

Electromagnetism is a theory of an electric field $E_i(x^0, x^1, x^2, x^3)$ and a magnetic field $B_i(x^0, x^1, x^2, x^3)$ (x^0 is time and this relativistic notation is used below in this book). Since i takes a value $1, 2$, or 3, the electronic and magnetic fields are "vector" fields. These are described by a relativistic 4-dimensional vector field $A_\mu(x)$ ($\mu = 0, 1, 2, 3$) which is more fundamental and called a "gauge field." The relation to the electric and magnetic fields is written through the "gauge field strength" $F_{\mu\nu}(x)$, defined as[1]

$$F_{\mu\nu}(x) = \partial_\mu A_\nu(x) - \partial_\nu A_\mu(x), \tag{2.1}$$

$$E_i(x) = F_{0i}(x), \quad B_i(x) = \frac{1}{2}\epsilon_{ijk}F^{jk}(x). \tag{2.2}$$

Here, ϵ_{ijk} is a tensor with completely antisymmetric indices (i, j, k), and $\epsilon_{123} = \epsilon_{231} = \epsilon_{312} = 1$, $\epsilon_{132} = \epsilon_{321} = \epsilon_{213} = -1$. ("Tensors" have multiple indices, and the number of the indices is called a "rank" of the tensor.) The reason why we write the electromagnetic fields in this manner explicitly is that, as I will mention later, it relates to a considerably important principle in elementary particle physics, called gauge symmetry.

At any rate, in the case of electromagnetism, both the electric field and the magnetic field are written by a single vector field $A_\mu(x)$. The particle corresponding to this $A_\mu(x)$ is nothing but the light. In order to emphasize its particle nature, it is usually called a photon.

2.1.2 Equation of Motion, Quantization and Elementary Particles

In the case of the water surface, it is known that this field $h(t, x, y)$ satisfies a certain differential equation of (t, x, y). Let us call it an "equation of motion" of a field theory of the field h.

In mechanics, especially in analytical mechanics, when an "action" is given, its equation of motion is derived by a variational principle. It is entirely the same in field theories. A certain action written by fields is given as a basis, a field equation of a motion is derived once we apply the variational principle to the action.[2]

---

[1]In this book, Einstein's convention is used, namely, we will make it a rule to sum subscripts appearing twice. And ∂_μ is an abbreviation of a partial differential operator $\partial/\partial x^\mu$. Moreover, we employ the unit system where velocity of light is $c = 1$, in this book.

[2]The variational principle is a principle deriving equations of motion from actions. In the present case of field theories, the situation is the same as that of an analytical mechanics where an equation of motion is derived from an action (a Lagrangian). Suppose an action is given. The action is written by action variables (which are fields in the case of field theories, and are coordinate $x(t)$ in the case of the analytic mechanics). Let's require that the action itself doesn't change by any infinitesimal change of the action variables (this is called "variational principle"). It gives a condition for the action variables to satisfy, and it is nothing but the equation of motion. In the case of the action $S[A_\mu]$ described by the field $A_\mu(x)$, the equation of motion is written as $\delta S/\delta A_\mu = 0$. This means an infinitesimal change of the action must be zero. Here, we have derived the equation of a motion

For instance, in the previous example of electromagnetism, the action is given as

$$S = \int d^4x \, \frac{1}{4} F_{\mu\nu}(x) F^{\mu\nu}(x). \tag{2.3}$$

By the variational principle, an equation of motion

$$\partial^\mu F_{\mu\rho}(x) = 0 \tag{2.4}$$

is derived. This is equivalent indeed to the well-known Maxwell's equation. For instance, Gauss law $\partial_i E_i = 0$ is derived for a component $\rho = 0$.[3]

The standard model which describes almost all physics of known particles consistently is basically written by an action like this. Actions give a definition of physical theories, and they consist of informations such as species of fields appearing in the theory, dimensions of spacetime on which the theory is defined, and forms and strengths of interactions of the fields. A unique action describes all matters and interactions in our whole world – which is quite amazing, and is a monumental achievement in modern physics.

The important point of field theories is that each field corresponds to each kind of particles. Precisely, this correspondence is called "second quantization." Small excited waves of a field appearing in an action is "quantized," and they become particles which are countable (one particle, two particles, ... This is called an elementary excitation), and represent particles appearing in the physical theory.[4]

Let us see how this works in more details. It is easy to imagine that the water surfaces have a generic solution which is a traveling plane wave. The simplest form should be

$$h(t, x, y) = a \cos[k_0 x^0 + k_1 x^1]. \tag{2.5}$$

Here a, is a very small constant which is the height of the wave, and x^0 is time, while x^1 means a certain direction of the (x, y) plane. This wave solution of the water surface equation of motion travels in the direction x^1. I do not write the explicit equation here, but the equation will be presented in the next section for a so-called scalar field. In a similar manner, for the electromagnetism, we have a solution for

---

with respect to the gauge field $A_\mu(x)$, and in the next section let's have a look at a derivation of an equation of motion for a little bit easier example of a non-vector $\phi(x)$.

[3] For bringing the indices up or down, we use a metric tensor for a flat spacetime (due to special relativity). The component of this tensor $\eta_{\mu\nu}$ is $\eta_{00} = -1$, $\eta_{11} = \eta_{22} = \eta_{33} = 1$, while the other components are zero. For instance, the Maxwell's equation $\eta^{\mu\nu} \partial_\mu F_{\nu\rho} = 0$ is interpreted with summed indices for μ and ν. Here $\eta^{\mu\nu}$ is an inverse matrix of $\eta_{\mu\nu}$, and as you know from the components just mentioned above, the actual form of the matrix is just the same.

[4] About how you may obtain a quantum mechanical description of scattering of particles from this action, you would consult an explanation of "a perturbation theory" in Sect. 2.4.

the equation of motion (2.4) which looks like a plane wave,

$$A_\mu(x) = a_\mu(k) \cos[k_\mu x^\mu], \tag{2.6}$$

when the constant $a_\mu(k)$ satisfies a constraint $k^\mu a_\mu(k) = 0$. This is an electromagnetic wave.

If we follow the "quantization" procedure, we reach an interpretation that this $a_\mu(k)$ indeed represents a photon. This photon has a momentum k_μ (this is a four-momentum notation in special relativity), and freely propagates in space. For any computation of scattering process of elementary particles, we first need to specify the incoming and outgoing states of the particles, which are indeed freely propagating in asymptotic space. The plane waves are identified as the asymptotic states of the elementary particles, and the state is specified by this $a_\mu(k)$ for photons. The quantization provides how you count the number of photons in the asymptotic states. This is made possible by upgrading the normal number $a_\mu(k)$ (or rather to say a function of k) to a "quantum operator." I will not go into the details, but the essence is that the state is specified by a product of the "operator" $a_\mu(k)$, and this is nothing but the way how we specify the number of the elementary particles at the asymptotic states, in computing the scattering processes in quantum field theories.

In reading the following of this book, the readers are just asked to keep in mind that each field in a quantum field theory has a corresponding elementary particles, and the quantum field theory has an action, and the standard model of elementary particles is written entirely in that manner, with given fields and given action.

2.1.3 Problems in Elementary Particle Physics

The standard model of elementary particles is written as a quantum field theory. It consists of a set of fields (each of which corresponds to a species of elementary particles), and a single unique action written by the fields. Techniques of quantum field theories can define any scattering (interaction) of elementary particles in a concrete mathematical manner, so, once the action is given, we can compute any scattering of elementary particles. And, this procedure of description of elementary particles and their interaction has been extremely successful. In fact, so far, almost all of particle experiments are consistent with the standard model of elementary particles.

At this stage, you may think that, okay, there is no need to improve the standard model, as it describes all known experiments. This is not the case. The standard model, or rather to say, particle theory, has serious problems. Superstring theory and D-branes, in addition to solitons, are for resolving those problems, which is a theme of this book. In this subsection, we just list the problems which will be discussed in later sections and chapters in this book.

Before listing the serious problems, let me provide two additional issues on the standard model of elementary particles. It is of course important to find an answer to these problems, but will not be treated in this book. First one is about "Higgs particle." A field corresponding to this elementary particle is included already in the standard model, but this particle has never been observed. It is expected that this particle is observed in the modern experiments using a particle accelerator called "Large Hadron Collider (LHC)." Once this particle is observed there, we will obtain information on how it interacts with other elementary particles, which will test the standard model of elementary particles. See Sect. 2.4.1 for more details. The second issue is the mass of neutrinos. In the standard model, elementary particles called neutrinos are supposed to be massless. However, recent experiments revealed that the masses are not zero, which says that the standard model needs a modification. Experiments ongoing will provide us with more information on the neutrino sector of the standard model, which is important for what kind of modification we need for improving the standard model.

These two issues are of course important, but a bit different in nature from the serious problems which are discussed in this book. Here is a list of the serious problems of the particle physics. The theme of this book is how these serious problems may be resolved once we consider superstring theory and D-branes. There are four questions:

- *Quantum gravity*: Quantizing gravity?
  The standard model is written by a quantum field theory, where fields are quantized to provide a particle picture. Now, Einstein's general relativity, which is the theory of gravity, is also written by a field, which is the gravitational field. But this gravitational field is not included in the standard model. We all know that gravity is coupled to any elementary particle physics, so the gravitational field should be included to the model and should be quantized as well as the fields of the elementary particles included in the standard model. However, there exists a theoretical difficulty in quantizing the gravitational field. There appears infinities in calculations which cannot be removed. We do not know how to quantize the gravity, by using the standard techniques of quantum field theories.

  This is the reason why we are interested in superstring theory. As we will see in Sect. 3.2.2, string theory naturally includes gravity and the gravitational field, and furthermore, string theory is expected not to provide any infinity, as we saw in Sect. 1.1.

  The most interesting object in gravity theory is black holes. At the center of the black holes, there are singularities where all physical laws are invalid. Quantum treatment of gravity is expected to resolve this issue, and thus the quantization of gravity is quite important. Quantum nature of black holes appears when one try to count number of quantum states of the black holes. Superstring theory and D-branes provide a way to give this number, as we will see in Sect. 6.3. Therefore, a first step toward finding a consistent quantization of gravity is made in superstring theory and D-branes.

  Gravity is intimately related to cosmology of our universe. Recent progress in observation of the cosmic microwave background revealed interesting properties of the early universe, and we start understanding that there was an era of

rapid inflation of the universe: inflationary cosmology. Then, how naturally can
we implement this inflationary expansion of the universe in our field theory
models of cosmology? And how quantization of gravity is related to this, at the
early universe? These are important questions in cosmology related to quantum
gravity, and we will see in Sect. 6.2 that superstring theory and D-branes provide
a simple interesting setup for the inflationary cosmology.

- *Arbitrariness problem*: Arbitrary interaction strengths among particles?
  In the standard model, the action is given, but there are plenty of room for
  modifying the standard model. For example, the electric charge of electrons
  can be modified by hand if you like. To reproduce the observed electric charge
  of electrons, we need to choose one number. But the standard model does not
  explain why the number is that. In the same manner, the masses of the elementary
  particles are arbitrary constant numbers in the model, and constants specifying
  all the interactions among particles are again arbitrary numbers. Why the nature
  chose such values of the interaction strengths and masses? This is called an
  arbitrariness problem.
     A famous attempt to resolve this issue a bit is a trial called "grand unification."
  The unification means gluing and combining various interactions in the standard
  model to a unique one, and in this way, we can reduce the number of arbitrary
  constants. At present, there is no direct experimental proof for this grand
  unification, but solitons are very important for a direct observation. Details are
  described in Sect. 2.4.1.
     A more radical unification can be made by superstring theory. In string theory,
  particles are replaced by oscillations of strings as will be described in Sect. 3.2.2,
  and resultantly, some of the interaction strengths are unified. This was briefly
  explained in Sect. 1.1. A modern version of this interaction unification using
  D-branes are described in Sect. 6.1.6. There, you will find that the interaction
  strengths in the standard model can be understood in a geometrical manner in
  higher dimensions in superstring theory.

- *Hierarchy problem*: Why gravity is so weak?
  One of the reasons why the gravity is not included in the standard model is,
  besides the problem of the quantization of gravity, that it is very weak compared
  to the other forces and so can be ignored at a first approximation. All the other
  forces are included in the standard model, for example the electromagnetic
  forces. So, the question is: why the gravity only is so weak? Is there any
  qualitative difference in theoretical formulation, between the gravitational forces
  and the other forces? There is apparently a hierarchical structure in the interaction
  strengths of the forces, this is the hierarchy problem of gravity.
     In Sect. 6.1, you will see that a novel scenario "braneworld" can solve this
  hierarchy problem. The braneworld scenario is the hypothesis that we are living
  on a hypersurface in higher-dimensional space. Only gravity can propagate out
  of the hypersurface, that is the origin of a possible solution of the gravitational
  hierarchy problem. The novel perspective of the braneworld scenario came out

of the physics of D-branes. The interesting physics behind this new perspective is already encoded in the physics of solitons, and you will see in Sect. 3.1 (in particular Sect. 3.1.4) how you can live only on such hypersurfaces, in terms of field theories. The important concept "solitons" are necessary to understand the idea, and solitons are explained in the following subsections in this chapter.

* *Quark confinement problem*: Why there is no observation of a single quark?
As is mentioned, all fields in the standard model have their corresponding elementary particles. This is also the case for quarks. We know that protons and neutrons are made of quarks, that is how almost all the hadrons in our world are understood and classified, as quark bound states. However, interestingly, no one has observed a single quark in experiments; what we see are only bound states of quarks. Why is this? And how is this realized? This problem is called "quark confinement problem" which is one of the serious problems in the standard model of elementary particles.

From the next subsection, we will see that in most of field theory models we can find "solitons" which are special solutions of equations of motion. And interestingly, you will see that solitons are expected to play an important role in solving the problem of the quark confinement, via so-called "duality" symmetry. It will be explained at the end of this chapter, Sect. 2.4.2.

D-branes in superstring theory provides us with a fascinating new idea called "holography" or "gauge/gravity correspondence." This new mathematical tool found in D-brane physics in superstring theory shows that in a certain limit the physics of quarks can be mapped to a physics of hypothetical gravity in higher dimensions. Using this new correspondence, once can compute interactions between quarks, to show the confinement. The new technology is completely based on the D-branes in superstring theory, which you can learn in Sect. 6.4.

The final goal of the elementary particle physics should be to find a solution to these serious questions. The excitement of researchers studying superstring theory and D-branes is mainly the fact that this theory can solve these questions, and at least at some level, some are solved, as you will see in this book. The notion of solitons in field theories, and that of D-branes in superstring theory, are related and provide a new paradigm for possible solution of these serious questions.

From the next section, let us start seeing how the theoretical notions such as solitons and D-branes are introduced, to find a novel possibility of resolving these problems in elementary particle physics.

2.2 What is Soliton?

To begin with, what kind of physics does the word "soliton" indicate? Soliton is a certain kind of objects considered in field theories in physics. I will introduce the definition of the solitons, and explain the field theories which are the basis for the solitons.

2.2.1 Field Theory and Soliton

The word "soliton" can be split into "solit-" and "-on". The latter part "-on" is often used for particles in physics. For example, electrons are particles carrying electricity. This usage is generic: particles transmitting gravity are called gravitons, and particles transmitting electromagnetic force are photons, and there are many other examples. Then, the problem is, what does the part "solit-" which should show a characteristic of the particles mean? This comes from a word "solitary". That is, solitons are solitary particles.

To see how the solitariness can be defined in physics, it is appropriate to refer to the way how the solitons were found, which I will tell you in the following. In 1834, John Scott Russell, who wondered at a canal in Edinburgh, found that waves on the surface of the Union Canal in Scotland were moving without breaking up for miles. In a common sense of water waves, for example, when you fall a stone into a still pond, waves form periodic co-axial circles origined at the point of the fall, and gradually expand and disappear finally. However, in the case of the waves of this canal, waves were solitary and kept moving without changing their height and their velocity, for miles. This shows that the wave behaves as if it were one particle ("-on"). Particle could solitarily keep traveling without any sudden disappearance. The waves of this canal correspond to Tsunami on sea surface, and they also share a property of slipping through each other in collision of two waves of the same kind. This phenomenon shows solitariness of waves.

The origin of the word "soliton" is due to these characteristics, while modern physics is described by mathematics called field theories, and solitons are the objects defined there. By using the words of the field theories, solitons in physics are "solutions of equation of motion of the field theory with their energy localized." In order to explain this meaning, at first, I will explain the notion of fields and field theories, by using this phenomena of the water surface.

We have seen previously that the water surface is described by a field $h(t, x, y)$ which is subject to an equation of motion, a differential equation. As a solution of it, there exists a soliton solution, namely, a solitary wave solution. Since this field theory of the water surface is not a "quantum field theory" which respects quantum mechanics, the solution is called a "classical solution."[5] Figure 2.1 shows a configuration of a solitary wave at a certain time.

Any water wave has a traveling direction, and the "height field" h of the simplest solution doesn't depend on the coordinates transverse to the traveling direction. Therefore, for instance, in the case of a wave traveling along the direction of x, $h(t, x, y)$ is independent of y, and is a function of only (t, x). The energy of this solution is localized at a certain place in the x direction for a fixed certain time. In this sense, water waves are solitons.

---

[5]In this way, solitons are introduced as classical solutions without a concern on quantum effects. However, as we will see later, solitons play an important role in quantum field theories.

$h(t = t_0, x, y)$

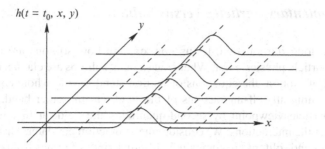

Fig. 2.1 A configuration of a solitary wave on the water surface at a certain time $t = t_0$. The wave travels toward the direction of x

However, as you see in Fig. 2.1, the energy of this soliton solution is not localized in y. The meaning of "localized" is loose, like this. Speaking with another expressions, this water wave is a solution of a partial differential equation whose variables are only (t, x) from the first place. Since there is no dependence on the direction y, we can put the differential with respect to y to zero in the partial differential equation, and you don't have to consider y from the first place. In the previous view, the theory is a field theory in 3-dimensions of (t, x, y). It is often called $2 + 1$ dimensions by writing the time direction separately. However, if there is no y at the first stage, this is a field theory in $1 + 1$ dimensions, and the soliton is entirely localized in the space x.

Now you would understand that solitons are solutions of equations of motion in field theories and also their energy is localized even if time passes. The definition of this word "soliton" differs in subjects, a little. For instance, in mathematics, there are cases in which solitons have another requirement that they are solitary that solitons travel through each other, as well as the previous conditions. On the other hand, in elementary particle physics the condition of this solitariness is usually not required, and just classical solutions having localized energy are called solitons. In the rest of this book, I use the term "solitons" in that sense.

String theory, the main subject of this book, is closely related with elementary particle physics, which uses field theories as its basis. Therefore, solitons, that is, solutions of equations of the motion in field theories, are as a matter of course very important. I will have a description of their importance over showing examples in Sect. 2.4 (and Sect. 3.1), but beforehand, here I briefly explain the relation and the present situation of solitons and string theory. To tell the truth, ultimate and fundamental action of the field theory of strings has not been found yet. However, it has been gradually found what kind of properties the action should possess. It is about solitons which should exist in string theory – The solitons are the very D-branes! A revolution has occurred in string theory, when this fact was found in 1995. In explosive developments after that, the D-branes which are solitons of string theory play a main role, furthermore, they give a vast influence on not only string theory but also various theories of physics.

2.2.2   Elementary Particles Versus Solitons

Now, after a long introduction, let us now explain how solitons are derived in elementary particle physics. First, When you say that solitons are classical solutions of equations of motion, the equations of motion stand for the whole equations of motion containing all self-interactions of the field. On the other hand, when we derive the particle viewpoint by second-quantizing the field, at first we usually don't consider the interactions: We consider the interaction part later. That is, in this sense particles and solitons are completely different kinds of things from each other. Both of them are objects derived from field theories and their energy is localized. However, in fact properties of elementary particles and that of solitons are different, and as a result, they play different roles.

Let us take the previous analogy of the water waves to see this difference. Suppose that surface of the water is rippled. When the height is small in such a case as the ripples, since as the field h is small, when the equation of motion is written by a polynomial expression of h (and its derivatives ∂h, $(\partial)^2 h$ and so on), we can use an approximation treating only terms linear in h. Namely, we can neglect the higher order terms such as h^2 and h^3, when h is small. These small ripples correspond to elementary particles. Small waves are given point-particle-like picture by quantum mechanics, and behavior as elementary particles. On the other hand, when h gets bigger, we must not neglect the terms of higher orders in h. Then we have to solve the whole equations of a motion at once. Apparently, this higher terms are interpreted as "self-interactions of h." As h gets bigger and bigger at a certain location, the field h around the location is influenced. The higher terms stand for the influence. As a result of the interaction with neighbors without using the approximation of small h, when the field h eventually takes a certain shape and keeps it while moving, it is called a soliton.

In standard model of elementary particles, for each of all the fundamental particles found ever, such as electrons, quarks, and photons, a corresponding field is introduced, and the fields are treated as particles by the quantization. Then what is the relation between solitons and the elementary particles? Though solitons have localized energy, the field at the localized place takes a special configuration which is far away from a "vacuum solution." The vacuum solution is a configuration with a constant or often vanishing field, and stands for a state of completely nothing (say, for the water waves, a state $h = 0$, at which the surface of the water is quite still enough and there is no ripple, corresponds to the vacuum solution). We can say that, at the location of the soliton where the field is locally far away from the vacuum solution, elementary particles obtained by the quantization of the field moves collectively. Namely, many elementary excitations of the field condense. This means that you can regard that solitons are generated by the collective motion of many elementary particles. This is obvious in the previous example of the water surface. A big solitary wave which is soliton-like is a big deformation from the flat water surface, and it is made by piles of small ripples. This image is shown in Fig. 2.2.

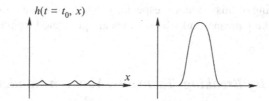

Fig. 2.2 *Left*: some ripples are interspersed. These correspond to elementary particles. *Right*: A soliton as a big solitary wave. Many particles group and take a special configuration to move collectively. Here, the appearance of the water surface at time $t = t_0$ is shown. There is no y-dependence in the field, and only the x direction is shown

Moreover, this soliton collectively move as a group and it looks like a particle by itself. Even though we call it particle-like, since a lot of elementary particles move as a collective motion, the mass of the soliton is quite heavy compared with the elementary particles. A distribution of the masses of various states appearing in a theory is called a "spectrum." Given a field theory, we can say that the particle spectrum include two categories, namely, elementary particles obtained by field quantization, and solitons.

What is the meaning of the existence of solitons in field theories which look like one heavy particle? In fact this question is closely related with what kinds of principles describe actions of field theories in elementary particle physics. The principle is "symmetry" which we describe in the next section. Once we make clear the relation between solitons and symmetries, it will naturally become clear what standpoint solitons are on in physics theories. And possibilities of solitons – a possibility that solitons are objects which can be observed, and a possibility to solve logical difficulties in field theories – appear from there. Let us take a look at these in Sects. 2.3 and 2.4, one by one.

2.3 Symmetries in Field Theories and Solitons

The standard model of elementary particles is written based on a certain symmetry. If I summarize a relation between the symmetry and solitons in one sentence, it reads as "Solitons are generated when a symmetry of the field theory is spontaneously broken." Since this sentence shows in what situation the solitons appear in elementary particle physics, this is very important in physics of solitons.[6] In this section, I will

---

[6]Some solitons often appear without any relation to symmetries. (Say, the example of the water waves.) In this book, we mainly treat solitons related with symmetries, because we consider solitons of elementary particle physics and string theory. All solitons which are introduced in Sects. 2.3 and 2.4 are of that type, and they are called topological solitons.

explain the meaning of this sentence, especially, what the symmetries are, what the spontaneous breaking means, and why solitons are generated from that.

2.3.1 Symmetry Breaking, Tachyons, Vacuum Condensation

In the standard model of elementary particles, the mechanism of "symmetry braking" is used essentially. Since the standard models of elementary particles is, unfortunately, very complicated and it is hard to tell the essence of how solitons are derived, let us take a look at a relation between solitons and symmetries by using a toy model.[7] The relation with the standard model will be mentioned later.

The previous solitary waves on the water surface are solitons having energy localized in the direction x. A typical example of a field theory which gives localized solitons in one-dimensional space in the same manner, is a theory called $1 + 1$ dimensional ϕ^4 model. This model has a certain simple symmetry, and solitons appear by its breaking. In order to break the symmetry, a condition is derived that at first this theory, as an appearance, must have tachyons which are particles of faster-than-light speed. After the symmetry breaking, these tachyons disappear. This is called "vacuum condensation," and it is a very important element of elementary particle physics. Here let us take a view around these physics by using the ϕ^4 model.

The ϕ^4 model is a theory of a field $\phi(x^0, x^1)$, and its action is written as

$$S = \int dx^0 dx^1 \left[\frac{1}{2} (\partial_0 \phi)^2 - \frac{1}{2} (\partial_1 \phi)^2 - V(\phi) \right]. \tag{2.7}$$

Here, we give the potential term $V(\phi)$ as

$$V(\phi) = \frac{1}{4} \lambda (\phi^2 + m^2/\lambda)^2. \tag{2.8}$$

m and λ are constant parameters. Then this action has the following symmetry, that is, the action is invariant under the following symmetry transformation,

$$\phi(x) \quad \leftrightarrow \quad -\phi(x) \tag{2.9}$$

because the action is all written by terms of even powers of ϕ. This is the symmetry of our concern.

---

[7] Toy models are various mathematical models (in particular, field theories) which physicists often use in favor. Reasons of actual phenomena of physics have varieties, and furthermore, are complicatedly involved each other. Therefore, we consider a hypothetical simple physics theory (a toy model) at first and examine its properties, and then by comparing them with the actual phenomena, we learn principles hidden behind the actual phenomena – this is our strategy.

The "potential" which appears in the action is a generalized version of a potential of mechanics, to field theories (for example, when an object is put at a heigh place on the earth, it has a gravitational potential).[8] What is the physical meaning of the parameters contained in the potential (2.8)? The parameter λ is a coefficient in front of the term ϕ^4, it is a "coupling constant" which shows the magnitude of the interaction in this theory. This is the reason why we call this theory as the ϕ^4 model. In addition, the parameter m is a mass of the particles coming from the field ϕ. Let us explain why. First, we shall derive an equation of motion by the variational principle from the action (2.7). Basically, we differentiate the quantity in the integration in the action (it is a Lagrangian) by the field ϕ. As for the term $\partial_1\phi\partial_1\phi$ containing the differentiation of ϕ with respect to x, let us use a technic: first we partially integrate it and rewrite it as $-\phi\partial_1\partial_1\phi$, and then we differentiate ϕ which is not act by the differentiation by x. Then finally, as an equation of motion we have

$$- (\partial_0)^2\phi + (\partial_1)^2\phi - m^2\phi - \lambda\phi^3 = 0. \tag{2.11}$$

For now, we would like to consider the mass of the particle, we neglect the interaction term $\lambda\phi^3$. Then, this equation is "linearized" (that is, it has only terms linear in ϕ) and has the following wave-type solution:

$$\phi = (const.) \cdot \cos[k_0x^0 + k_1x^1 + (const.)], \tag{2.12}$$

$$k_0 = \sqrt{(k_1)^2 + m^2}. \tag{2.13}$$

Here, since (k_0, k_1) is a relativistic momentum, k_0 is the energy and k_1 is the momentum. Namely, (2.13) is a famous relation of relativistic energy of particles[9] and m is the mass of the particle. This particle is interpreted as an elementary particle derived from a second quantization of the field $\phi(x)$.

At this stage, in this ϕ^4 model, suppose that m^2 is a negative number. (That is, m is considered as a purely imaginary number.) In fact, this allows a soliton solution in the theory. We will explain it concretely later, but before that, let us examine in detail what is the meaning of setting m^2 negative. It is closely related with how the symmetry of the theory is broken.

When m^2 is negative, the velocity k_1/k_0 of this particle is beyond the speed of light 1, it becomes what is called "tachyon." The word tachyon sounds like a

---

[8]In the same manner as the mechanics, Hamiltonian can be derived from the action, and it stands for energy of the field:

$$H = \int dx^1 \left[\frac{1}{2}(\partial_0\phi)^2 + \frac{1}{2}(\partial_1\phi)^2 + V(\phi)\right]. \tag{2.10}$$

This expression shows that the potential $V(\phi)$ by itself directly contributes to the energy.

[9]In the case of momentum $k_1 = 0$, this equation is equivalent to $k_0 = m$. Bringing back the light velocity $c = 1$, it becomes the famous $E = mc^2$.

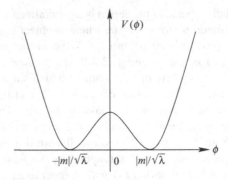

Fig. 2.3 The potential for negative m^2 in the $\phi = 0$ model. The point $\phi = 0$ is placed at the top of the mountain of the unstable potential, and the actual vacua are placed at the right of it or at the left

scientific fiction. Using a tachyon makes it possible to communicate with a speed faster than light, so, does it mean a contradiction to relativity? the answer is no. In a certain field theory, even if it contains a tachyon field, it has no problem against relativity. This is because a "vacuum condensation" which will be described below will help the theory from the contradiction.

The potential $V(\phi)$ is called a double well potential, and two values $\phi = \pm|m|/\sqrt{\lambda}$ of the field are its bottoms. Take a look at Fig. 2.3. The potential is concave around the point $\phi = 0$. That is, the theory is unstable at $\phi = 0$. That the potential is concave corresponds to the negative m^2, and as a result it appears that a tachyon comes out. Since any theory generally favors stable states with lower energy, ϕ falls and rolls down to the bottom. The value of the field at this bottom stands for the "vacuum" of this field theory. A vacuum is a state with no particle in the sense of quantum field theory. The phenomenon of rolling down to the bottom of potential is called "vacuum condensation." At the vacuum where the potential is convex and thus stable, if we expand the field around it, the mass squared of the elementary particles which appear there is positive, properly.[10] This would be clear when you see the figure. Namely, the tachyon does not actually exist, and we have just misunderstood as if the tachyon could exist because we look at all things just around the unstable $\phi = 0$.

By the way, in this vacuum condensation mechanism, it is important that the symmetry the theory originally had is spontaneously broken. The ϕ^4 model has the symmetry of changing the sign of the field ϕ as $\phi \to -\phi$. However, after one among the true vacua $\phi = \pm|m|/\sqrt{\lambda}$ is chosen, the symmetry is broken (see Fig. 2.4). This is called a "spontaneous symmetry breaking."

The situation like this is not only in this model, and in fact, even in the standard model of elementary particles, the same vacuum condensation occurs. In

---

[10]Writing $\phi = |m|/\sqrt{\lambda} + \delta\phi$ and substituting this to the action, you get the mass as a coefficient of $(\delta\phi)^2$.

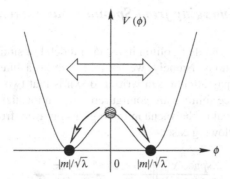

Fig. 2.4 Spontaneous symmetry breaking. The system changes from the unstable state $\phi = 0$ to the stable vacua $\phi = \pm|m|/\sqrt{\lambda}$ (the "vacuum condensation"). The *horizontal arrow* stands for the symmetry $\phi \to -\phi$ of the system, and after the vacuum condensation this symmetry is broken spontaneously

the standard model, the field ϕ is called "Higgs field." The Higgs field has the same kind of potential which generates the vacuum condensation. In the standard model, there exists a symmetry larger than the ϕ^4 model, which breaks due to a vacuum condensation. As we saw in the example of the electromagnetism contained in the standard model, the electromagnetic field, namely, the strength $F_{\mu\nu}$ is written by the gauge field $A_\mu(x)$. As can be understood in (2.1), it is invariant under the following transformation with an arbitrary function $\Lambda(x)$,

$$A_\mu(x) \to A_\mu(x) + \partial_\mu \Lambda(x). \qquad (2.14)$$

Therefore, the action (2.3) of electromagnetism is also invariant under the transformation. The invariance under an arbitrary function $\Lambda(x)$ like this is called a "gauge symmetry." Since this symmetry is for an arbitrary function $\Lambda(x)$, the constraint to the theory is very strong, therefore it can be a ruling principle for writing down actions. In fact, the standard model of elementary particles has a gauge symmetry generalized in such a way that it contains the gauge symmetry of this electromagnetism. However, since the symmetry like that is not seen in nature, the standard model is technically made to break spontaneously the large gauge symmetry by the Higgs field. After the spontaneous symmetry breaking, the gauge fields for the broken gauge symmetry acquire a mass. This mechanism is called "Higgs mechanism."[11]

When a symmetry of a field theory is broken spontaneously by a vacuum condensation, solitons often come up with it. The ϕ^4 model we saw here is a typical example of it. Next, let us see its solitons concretely.

---

[11]S. Glashow, S. Weinberg, and A. Salam received a Nobel Prize in 1979 for their work on an application of this mechanism to the standard model.

2.3.2   Solitons Emerging from Spontaneous Symmetry Breaking

Let us write down concretely a soliton in the ϕ^4 model. For simplicity, we consider time-independent solitons. Namely, we assume ϕ to be a function of only x^1.[12] The points we must pay attention in writing down is that two stable vacua ($\phi = \pm|m|/\sqrt{\lambda}$) exist. Since solitons are configurations with localized energy, the value of the field should equal to the vacua in the regions far away from the solitons. So, let us consider the following case:

$$
\begin{array}{c|c|c}
\text{space } x^1 & -\infty & +\infty \\
\hline
\text{field } \phi(x^1) & -|m|/\sqrt{\lambda} & +|m|/\sqrt{\lambda}
\end{array}.
\tag{2.15}
$$

This is a configuration of ϕ moving from one vacuum to another vacuum at a certain point in x^1. On the transition, the field must climb up the potential hill, and $\partial_1\phi$ is not zero there since the value of ϕ moves so it contributes to the energy (see the Hamiltonian (2.10)). Around there the energy is localized, and a soliton exists there.

This configuration is a solution of the equation of motion and is stable. In order to see it, let us visualize this soliton pictorially. We now look at a space spanned by x^1, ϕ, and $V(\phi)$ (Fig. 2.5). The shape of a classical solution is like a chain lying on a galvanized sheet iron roof which is snaked. Here, the chain stands for $\phi(x^1)$ and extends from $x^1 = -\infty$ to $x^1 = +\infty$. The stable vacuum solution is a state of the chain put totally in a single ditch of the roof. In Fig. 2.5, it is shown with a bold dotted line. On the other hand, let us consider the soliton with the boundary condition (2.15) mentioned earlier. The corresponding chain should move from one ditch to another ditch at some point. This is shown in Fig. 2.6. When the chain passes the top of the swell of the roof, it is at a high place and earns $V(\phi)$ corresponding to the potential energy for the height. Furthermore, since it changes the place, it needs more length (The derivative $\partial_1\phi$ is nonzero). This point where the chain passes the swell is the location of the soliton, namely, the place where the energy is localized. At other places, the chain lies down at the bottom of the ditch, and the density of energy vanishes. By this visualization, it is clear that the solution like that exists and is stable.

It is also possible to get the soliton solution by solving the equation of motion concretely. To get a static classical solution, we may neglect time derivatives $\partial/\partial x^0$, and then the equation of motion (2.11) is

$$
(\partial_1)^2\phi - m^2\phi - \lambda\phi^3 = 0.
\tag{2.16}
$$

---

[12]In field theories, for solitons of classical field theories we often consider ones independent of time x^0, and they are called static solitons. In the case of relativistic physical systems, moving solitons are obtained by acting a Lorentz transformation of special relativity to the static solitons. That is, we can get a solution of a moving soliton by using the fact that for a moving observer even a static object could be seen as one moving in the opposite direction.

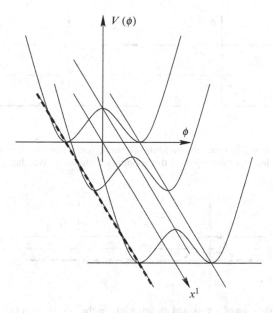

Fig. 2.5 A hypothetical space spanned by x, ϕ, and $V(\phi)$, is shown. The potential $V(\phi)$ is given as a surface which looks like a galvanized sheet iron roof, in this 3-dimensional space. A *bold dotted line* stands for one of the vacuum solutions $\phi = -|m|/\sqrt{\lambda}$. Since the vacuum solution does not depend on x, it linearly lies down at the bottom of the ditch

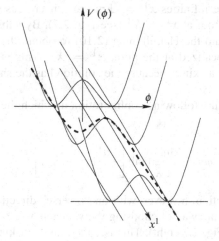

Fig. 2.6 A soliton solution. A chain (*bold dotted line*) once crosses over the swell of a galvanized sheet iron roof

The simplest solution is

$$\phi = \widehat{\phi}(x^1) = \frac{|m|}{\sqrt{\lambda}} \tanh\left[\frac{|m|}{\sqrt{2}}(x^1 - X^1)\right]. \tag{2.17}$$

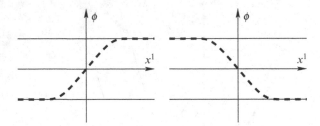

Fig. 2.7 A kink solution (*left*) and an anti-kink solution (*right*). Both of these are solitons, with only a difference of just opposite signs of the boundary conditions. We choose $X^1 = 0$ in the figure

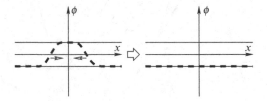

Fig. 2.8 A pair annihilation of a kink and an anti-kink. In the *left* figure, a kink and an anti-kink stay further each other. In the *right* figure they collide and annihilate and turn to get back the state of vacuum solution

Here, X^1 is an arbitrary real constant. As we expect, this solution satisfies the condition (2.15) at the infinities $x^1 = \pm\infty$, and it crosses across a swell of a galvanized sheet iron roof at $x^1 = X^1$ (see Fig. 2.7). By substituting this soliton configuration (2.17) into the Hamiltonian (2.10) to obtain the energy density, we exactly find energy localized at the point $x^1 = X^1$. This is a soliton placed at $x^1 = X^1$. We call this a "kink" because the solution has the shape given in the left figure of Fig. 2.7.

On the other hand, the following configuration in which the sign of the solution is flipped,

$$\phi = -\frac{|m|}{\sqrt{\lambda}} \tanh\left[\frac{|m|}{\sqrt{2}}(x^1 - X^1)\right], \tag{2.18}$$

is also a classical solution, however, with an "inverse" direction, compared to the previous one. That is, the way of choosing the vacua at $x^1 = -\infty$ and $x^1 = +\infty$ is taken flipped (see Fig. 2.7 right). This is called an "anti-kink." The distribution of the energy density is just the same as the kink. As we put the kink and the anti-kink far away from each other and then bring them closer gradually, they happen to pair-annihilate each other finally (Fig. 2.8). This is similar to a pair annihilation of a particle and an anti-particle. Note that the boundary conditions at the infinities of x do not change in the process of this pair annihilation. That is, the process of the pair annihilation of the solitons occurs locally. The locality of the pair annihilation is common to that of particles, and also in that sense, solitons behave like particles.

In this way, we saw that, from the concrete construction of the soliton solutions, they have localized energy and behave like particles, together with the anti-solitons. However, it appears that this soliton is in the 1-dimensional space and has nothing related with the 3-dimensional space we live in. Furthermore, if we try to put two solitons in this 1-dimensional space (without using the anti-soliton), we run into a problem: allowed configurations of solitons are only the case of solitons and anti-solitons put alternatively. These problems are resolved once we consider solitons in a little bit higher dimensions. Since solitons in not only one dimension but also two and three dimensions are considerably related with particle observations in elementary particle physics and cosmologies, in the next subsection, let us consider those solitons in higher dimensions.

2.3.3 Vortex: Soliton in Two Dimensions

The reason for the existence of the solitons in the ϕ^4 model is that there are two vacua in the theory. Namely, related with the symmetry breaking, the two field configurations $\phi = \pm|m|/\sqrt{\lambda}$ became the vacua. Here, note that there are two infinities $x^1 = \pm\infty$ in the 1-dimensional space on which the theory is defined. The reason why the soliton exists is that, as shown in Fig. 2.15, there are non-trivial ones among the maps which map the two infinities to the set consisting of the two points of the field vacua. In other words, the map

$$\{-\infty, +\infty\} \xrightarrow{\phi(x^1)} \left\{|m|/\sqrt{\lambda}, -|m|/\sqrt{\lambda}\right\} \tag{2.19}$$

has a degree of freedom to choose whether $\{-\infty, +\infty\}$ is mapped to the same vacuum or to different vacua for each.

This way of thinking can be applied not only to the one-dimensional case but also to two, three, or even higher dimensions. Here, as an example, we shall consider the two dimensions in detail. The example of the two dimensions gives us an example of countable solitons – one soliton, two soliton, etc., and moreover, it is closely related with creation/annihilation of D-branes in string theory which will be described in Sect. 5.3.

The 2-dimensional space is spanned by x^1 and x^2, and its asymptotic infinity is considered to have the shape of a circumference (see Fig. 2.9).[13] Let us consider a field theory on this 2-dimensional space. (This is a field theory on $2+1$-dimensional spacetime, if we count time. The example of the previous soliton is a $1+1$-dimensional spacetime.) As a generalization of the previous ϕ^4 model, suppose we

---

[13]In the polar coordinate $z = re^{i\theta}$, the asymptotic infinity corresponds to a limit $r \rightarrow \infty$. The asymptotic infinity of 2-dimensional space is characterized by $0 \le \theta < 2\pi$, and its shape is a circumference.

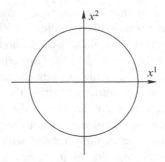

Fig. 2.9 Infinities of the 2-dimensional space spanned by x^1 to x^2 equal a circumference. Since an asymptotic infinite point can be determined by the direction along which one goes away from the origin, so the angle of the direction determines the point at the infinity. It is equivalent to determine one point on a circle

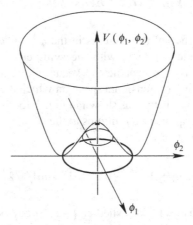

Fig. 2.10 The potential of the complex field. It can be seen that there is a rotational symmetry. The vacuum of the theory is the bottom of the potential, and it has the shape of a circumference (*bold line in the figure*)

have two real fields ϕ. This can be considered as a complex field, if we combine them as $\phi_1(x) + i\phi_2(x)$. And as a potential appearing in an action, we consider a generalization of the previous ϕ^4 model,

$$V(\phi_1, \phi_2) = \frac{1}{4}\lambda \left((\phi_1)^2 + (\phi_2)^2 + \frac{m^2}{\lambda} \right)^2 . \tag{2.20}$$

We assume that m^2 is negative as we set before. This potential is called "wine-bottle type," and I show the shape in Fig. 2.10. As you can find from this figure, this theory contains an interesting symmetry and vacuum. First, there is a symmetry:

$$\phi_1(x) + i\phi_2(x) \rightarrow e^{i\varphi} \left(\phi_1(x) + i\phi_2(x) \right) . \tag{2.21}$$

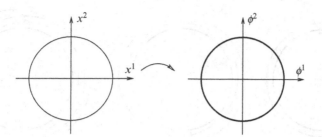

Fig. 2.11 Consider a map from a circumference to another circumference. The *left circumference* stands for the asymptotic infinity of the 2-dimensional space, and the *right one* stands for the vacuum of the field theory

This is a degree of freedom to change the phase of the complex field. φ is an arbitrary constant varying from 0 to 2π, and this symmetry is continuous. You can see that under this symmetry transformation the potential (2.20) is invariant.

This continuous symmetry is broken spontaneously, as before. The vacuum, namely the bottom of potential, can be found from (2.20) as an arbitrary pair (ϕ_1, ϕ_2) satisfying the equation

$$(\phi_1)^2 + (\phi_2)^2 = \frac{|m|^2}{\lambda}. \tag{2.22}$$

In this sense, the bottom of the wine bottle is the vacuum (see Fig. 2.10), and the shape of the vacuum is a circumference.

Now once that a certain pair (ϕ_1, ϕ_2) satisfying (2.22) is chosen, namely, one point on the circumference is chosen, the symmetry (2.21) disappears. This is a spontaneous symmetry breaking (Fig. 2.11).

Let us see solitons showing up in correspondence to this symmetry breaking. Since solitons have localized energy, the field at the asymptotic infinity far enough away from the solitons must be at the vacuum. The way a place in the asymptotic infinity corresponds to the point in the vacuum classifies the kinds of solitons. This is the same logic applied for the classification of the kink and the anti-kink previously. In the present case, the asymptotic infinity of the 2-dimensional space is given by one circumference. On the other hand, the vacuum is also given by one circumference. That is, we learn that classification of the maps from one circumference to another gives the solitons. This map can be easily understood by regarding the first circumference as a circular rubber band and the second circumference as a thick stick. Apparently, the maps are classified by how many times we wind the rubber band (see Fig. 2.12). As the winding has an orientation (there is an opposite winding), including the case of no winding (which corresponds to the vacuum), the windings turn out to be classified by an integer number. This is the "particle number" of the solitons. Negative integers correspond to the existence of anti-solitons whose number is its magnitude of the integer. Furthermore, if solitons and anti-solitons exist at various places, this number stands for the whole number of them (where we count the anti-solitons with minus sign). That

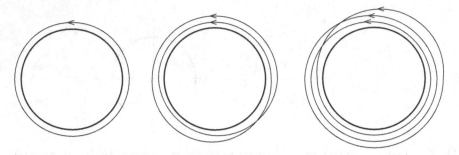

Fig. 2.12 Maps from a circumference to another are classified by winding numbers. In the *left*, the winding number is 1, and in the *middle*, the winding number is 2, while the *right* one has the winding number 3. *Arrows* are for showing the orientation of the winding. Opposite orientation of the *arrows* correspond to negative winding numbers. There is also a case with vanishing winding number

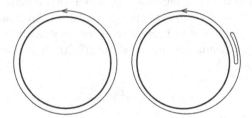

Fig. 2.13 The concept of the winding number is intact even when the rubber band is loose. In the *left*, a rubber band winds the circle tightly, while in the *right* a rubber band winds with a little folding. However, we consider that both give the same winding number

is, the solitons in this 2-dimensional space are countable, "one soliton, two solitons, $\cdots$" and we can consider states of multi-solitons. On this point, the situation is different from the previous ϕ^4 model, but common is the essential part concerning the conditions for the existence of the solitons, namely, the logic that solitons are classified by maps between the asymptotic infinity and the vacuum.

The integer label we saw here is called a homotopy group in mathematics. Homotopy is a property of invariants under continuous deformation of maps. For instance in the present case, even if a rubber band is loose or folded, the winding number is the same as that for a band which winds the stick tightly.[14] This agrees with the requirement that the number of elementary particles must conserve. That is, solitons behave as if they are particles (Fig. 2.13).

Solitons whose energy is localized in two dimensions like this example are called vortices. The integer labeling them is called vortex number. The vortex is an interesting soliton in 2-dimensional space. However, as a matter of fact, if we write down equations of a motion from the action containing the potential (2.20)

---

[14]Static soliton solutions solve the equations of a motion concretely, so they attain minimum energy configurations. And so they have the shape of the rubber tightly wound.

and solve the equations of a motion under the condition that the vortex number is not zero at the asymptotic infinity, then we will find that the total energy is divergent. As this shows the mass of the soliton is infinite, this is not a physically interesting solution at all.

To resolve this problem, combine electromagnetism with the theory of the complex field $\phi_1(x) + i\phi_2(x)$. Originally, the theory of the complex field has the symmetry (2.21) which changes a constant phase of the field. On the other hand, electromagnetism has also a symmetry (2.14) by a parameter of an arbitrary function. The symmetry of the combined theory is to perform (2.21) simultaneously with the gauge transformation (2.14) of electromagnetism,

$$\phi_1(x) + i\phi_2(x) \to e^{i\Lambda(x)}(\phi_1(x) + i\phi_2(x)), \qquad (2.23)$$

$$A_\mu(x) \to A_\mu(x) + \partial_\mu \Lambda(x). \qquad (2.24)$$

Note that here the constant parameter φ of the symmetry (2.21) for the complex field is upgraded to an arbitrary function $\Lambda(x)$. Now that the gauge symmetry of electromagnetism and the constant phase symmetry of the complex field are combined to a single gauge symmetry. Since also in this combined theory[15] the potential of the complex field is shaped like a wine bottle, a vacuum condensation occurs and then the gauge symmetry is broken spontaneously. This is the Higgs mechanism aforementioned. Though the analysis is omitted in this book, vortex soliton solutions exist as well in the combined theory, and what is more, the total energy of the soliton becomes finite due to the electromagnetic field. Interestingly, in this solution, a nonzero magnetic field $F_{12}(x)$ is concentrated around the center of soliton.[16] That is, solitons have magnetic charges.

2.3.4 Monopole: Soliton in Three Dimensions

By applying the example of the vortices in 2-dimensional space, we can consider solitons localized in 3-dimensional space, which is familiar with us. Probably you may easily guess what it will be, but let us write this down concretely. First of all, the asymptotic infinity of 3-dimensional space is given not by a circumference but by a spherical surface. And as a further generalization of the previous complex field, we consider three real fields ϕ_1, ϕ_2, and ϕ_3, and a potential

$$V(\phi_1, \phi_2, \phi_3) = \frac{1}{4}\lambda \left((\phi_1)^2 + (\phi_2)^2 + (\phi_3)^2 + \frac{m^2}{\lambda} \right)^2 \qquad (2.25)$$

---

[15]This model is called Abelian–Higgs model.

[16]In the electromagnetism of $2 + 1$ dimensional spacetime, the magnetic field has only one component F_{12}, as opposed to the familiar case of the $3 + 1$ dimensional spacetime. On the other hand, the electric field has two components of a vector: $F_{01} = E_1$, $F_{02} = E_2$. In electromagnetism of lower $1 + 1$ dimensions, there is only a single component $F_{01} = E_1$ of the electric field.

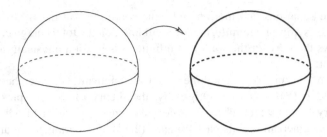

Fig. 2.14 The classification of maps from a sphere to another sphere gives us the number of monopoles. Though it is difficult to imagine a picture of windings, if one considers the case of winding by allowing it to crossing itself, one can find out that the winding number is given by an integer, just as in the case of the circumference

with $m^2 < 0$. Then the vacuum of this theory is given by

$$(\phi_1)^2 + (\phi_2)^2 + (\phi_3)^2 = \frac{|m|^2}{\lambda}, \tag{2.26}$$

which has the shape of a spherical surface. In this case, solitons are classified by maps from a sphere to another sphere (Fig. 2.14), and they are characterized by an integer which is the number of solitons, as in the case of the vortices. What is more, by combining a generalized electromagnetism (which will be explained in the following), the monopole is shown to have a magnetic charge at its core. This soliton solution is called monopoles (magnetic monopoles).[17]

Interestingly, the emergence of this monopole is deeply related with the standard model of elementary particles. The standard model of elementary particles is based on a field theory called "non-Abelian gauge theory" which is a generalization of electromagnetism. The gauge symmetry of the non-Abelian gauge theory is one of the most important principles in elementary particle physics, and this is a key point for how to measure the monopoles and how the elementary particle physics spreads beyond the standard model. Let us learn the non-Abelian gauge theory through the monopoles, below.

For the monopoles which are solitons in 3-dimensional space to appear, it is important that the vacuum of the field theory has the structure of a spherical surface. On the other hand, it is known that the monopoles would have infinite energy and become meaningless if the theory is not combined with an electromagnetism-like theory, as in the case of the vortices. So, let us consider what kinds of the electromagnetism we need to combine, from the viewpoint of the symmetry. Since in the case of the vortices the parameter of the symmetry transformation for $\phi_1(x), \phi_2(x)$ is a constant parameter φ, we have upgraded it to the arbitrary function $\Lambda(x)$ and could identify the symmetry with the gauge symmetry of the

---

[17]The existence of this solution was proved by G.'tHooft and A. Polyakov.

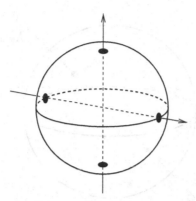

Fig. 2.15 When we perform a rotation around a certain axis and then a next rotation around another different axis, if we invert the order of the choice of the axes of the rotations, the result is different, even though we choose the same rotation angles

electromagnetism. On the other hand, in the case of the monopoles, the parameter of the symmetry transformation for $\phi_1(x), \phi_2(x), \phi_3(x)$ is not a single but three constant parameters. This is because this symmetry is a symmetry rotating the spherical surface, namely, a rotation symmetry in 3-dimensional space, and is specified by three angles (called Euler angles). When upgrading this to arbitrary functions, we have a gauge symmetry with three arbitrary functional parameters. That is, in this case, three gauge field will appear.

Furthermore, the rotational transformations with these three parameters are not mutually "commutative" operations. For instance, we consider, in a 3-dimensional space, a spherical surface which is the vacuum structure and two rotational axes there (Fig. 2.15). For example, performing a rotation by 30° around the first axis and then a rotation by 20° around the second axis, does not give the same result as performing the rotation by 20° around the second axis first and then the rotation by 30° around the first axis. This is expressed as these rotational operations being "not commutative." This time, the gauge symmetry possesses this non-commutative nature, it is not appropriate to bring three copies of the electromagnetism. This kind of generalization of the electromagnetism is called "non-Abelian gauge theory."

In the field theory of $\phi_1(x), \phi_2(x), \phi_3(x)$ combined with the non-Abelian gauge theory, the spontaneous breaking of the gauge symmetry by the vacuum condensation occurs, due to the potential of $\phi_1(x), \phi_2(x), \phi_3(x)$, as before (Higgs Mechanism). However, interestingly enough in this case, not all the symmetries are broken. Let us see the reason for this in Fig. 2.16. Suppose that a certain point (point X in the figure) on the spherical surface is chosen by the vacuum condensation. Then, although the symmetry rotating the whole spherical surface arbitrarily is broken, the symmetry rotating the spherical surface while keeping the chosen point fixed remains unbroken. Since this remaining symmetry does not change the circumference of equal distance from the chosen point, it is equivalent to the symmetry of the electromagnetism which we saw in the example of the vortex in 2-dimensional space. One out of three gauge symmetries remains unbroken.

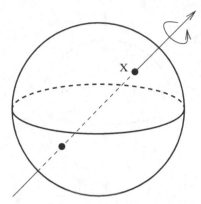

Fig. 2.16 The vacuum is a point on the spherical surface (point X). Once this point is chosen, it remains a rotational symmetry around the axis passing through point X and the center of the sphere

I hope now you can easily imagine the appearance of monopoles in the spontaneous symmetry braking of the non-Abelian gauge theory. In the part of the field theory of $\phi_1(x), \phi_2(x), \phi_3(x)$, the asymptotic condition at the infinity of 3 dimensional space generates the solitons. And, for this soliton, looking at the magnetic field of the electromagnetism remaining unbroken which is mentioned earlier, we can show that the magnetic field emanates from the center of the soliton (I will omit the calculation of this). This is the reason why the soliton in this 3-dimensional space is called a monopole, namely, a magnetic monopole.

Since the monopoles are localized in 3-dimensional space, it is the a particle-like object similar to elementary particles we know well. Therefore, it is possible that it can be observed in our 3-dimensional space we actually live in. This is related with the possibility to open a way for a new physics beyond the standard model of elementary particles. On the other hand, as monopoles appear, as well as particles, in the particle spectrum of field theories in 3+1-dimensional space time, there is a possibility to give us a new symmetry which may overcome difficulties in calculations in field theories. (This is called "duality" and is very important as being related with D-branes in string theory.) In this way, solitons play vivid roles in various aspects of field theories. We shall see these two possibilities in the following.

2.4   Importance of Soliton: Observation and Duality

2.4.1   Observing Soliton: Monopole in Grand Unified Theory

To see solitons, just go to a river or a seaside. Solitary waves created when a ship goes on the wave surface are solitons. The concept of solitons was found at a canal with the solitary waves on the water surface. In solid-state physics, actually there are solitons which can be observed. Here, in the elementary particle physics, which is closely related to string theory, I will give you an explanation of solitons there and their existence, and possibilities of their observation. The particle-like solitons I mentioned earlier are monopoles. How can they be observed, and what does it mean in elementary particle physics?

Monopoles appear according with the gauge symmetry breaking in non-Abelian gauge theories. On the other hand, the same kind of non-Abelian gauge symmetry breaking is used in the standard model of elementary particles. However, unfortunately, since in the standard model of elementary particles the structure of the vacuum is slightly different from the case of the monopoles written by ϕ_1, ϕ_2 and ϕ_3, solitons localized in the 3-dimensional space do not exist. This is unfortunate, but, there are more important monopoles. Those are monopoles appearing in "grand unified theories". I will explain it here. What are the grand unified theories beyond the standard model of elementary particles? And what is the meaning of the monopoles there?

The accuracy of the standard model of elementary particles has thoroughly proved by various accelerator experiments. It is perfect enough that no experimental result contradicts with the standard model.[18] However the standard model has a problem. The standard model contains a lot of constant parameters standing for the interactions, and the theory doesn't know why they take those values, and so they are given by hand at the step of writing the action. This is not satisfactory, and is a big problem in elementary particle physics. The mechanism determining such constants naturally is demanded. If it is found, we could predict, for example, the mass of the electron.

As one of attempts solving this problem, there is a field theory called "grand unified theory". This is written as a non-Abelian gauge theory in the same manner as the standard model, but the gauge symmetry is larger and contains the whole gauge symmetries of the standard model. And the attempt of the grand unified theory is to reproduce the whole standard model, by using the Higgs mechanism in this large gauge symmetry to break it to the gauge symmetries of the standard model. Anticipating the grand unified theory like this can give a relation among some parameters in the standard model, and can reduce the numbers of arbitrary parameters there.

If we assume this grand unified theory, the monopole, namely, the soliton associated with spontaneous symmetry breaking must appear. In fact, it is proved that a monopole must appear in a certain kind of grand unified theories.[19] That is, the monopole is the very evidence of the existence of the grand unified theory.

---

[18]Though it is supposed that the particle called neutrino is massless in the standard model, experimental results at the observatory Super Kamiokande in Gifu prefecture were reported that the neutrino has a mass. At the present, this is considered to be the only clear evidence deviating from the standard model.

[19]This is the case when the "group" defining the gauge symmetry of the grand unified theory is "semi-simple." The theories with a semi-simple group are simplified. Ideal grand unified theories should have fewer number of the parameters, and the case is a main subject of researchers.

Then, why hasn't this monopole been found in experiments before? It is because the monopole is considerably heavy. This is related to the fact that the energy scale of the gauge symmetry breaking of the grand unified theory is very high. The energy scale of the symmetry breaking corresponds, for example in the ϕ^4 model, to the difference between the value of potential $V(\phi = 0)$ at $\phi = 0$ and the value of potential $V = 0$. A picture is as follows: When a certain field at the vacuum (the bottom of potential) has the energy corresponding to the height of barrier of the potential, it can climb from vacuum at the bottom up to the top of the hill of potential. In the potential of the grand unified theory, the magnitude of this energy is set to be considerably high. This is from a theoretical requirement that the standard model should be included in it, here I will not describe it in detail (it will be explained in Chap. 6.) The most important is that the mass of monopole is large as this breaking energy (= the height of the potential) is high. As you can see in Fig. 2.6 of the kink solution of the ϕ^4 model, solitons climb over the potential barrier at least once. From this it is concluded that solitons must have the mass of order of the energy scale of the symmetry breaking.

The modern verification of elementary particle physics is mainly made by huge particle accelerators. Particle accelerator accelerates a lot of particles and gives them a large amount of energy and makes them collide. A pair of a new particle and its anti-particle (which has the same mass as the particle and has an opposite sign of charge) having the mass which amounts to the collision energy is pair-created at the accelerator, and we are able to learn a new physics by observing them. However, the mass of the monopoles expected in the grand unified theory is too large to reach even by the newest huge accelerator LHC which was activated in 2008. Therefore, to pair-create monopoles at huge accelerators is a task of extreme difficulty, and is not a realistic experimental subject.[20]

Then, can't we observe monopole? In truth, though monopoles of the grand unified theory cannot be created by mankind at accelerators, there must exist a lot of monopoles in the space of the universe. In the current standard model of the universe, the universe is supposed to begin with a big explosion called Big Bang. At the initial era of the universe it was extremely hot, but it was getting colder and colder as time went. That is, in the era close to the beginning of the universe, there must have been a period when the whole universe is hotter compared to the energy scale of the grand unified theory. Being hot means that the whole universe is full of energy, in other words, it was in the state at which the field ϕ can climb from the vacuum to the top of the hill of the potential. And as the universe expands and the temperature decreases, the spontaneous symmetry breaking occurs. As a result of this breaking, the monopoles are created. The creation mechanism of solitons like this is called Kibble mechanism. A lot of the monopoles of the grand unified theory were created in early universe and they must remain in the universe somewhere now.

---

[20]In the brane world which will be described in Chap. 6, the symmetry breaking energy of the grand unified theories might be brought much lower. If brane world is real, we might be able to produce solitons at accelerators.

How can we see the monopoles of the grand unified theory left over? Since these monopoles are related with the higher symmetry which is not present in the standard model of elementary particles, they create interactions which do not exist in the standard model, among particles. In this sense, if we find these interactions which we cannot interpret in the standard model, they would be the evidence of the existence of the monopoles. A typical example of the interaction is a decay of protons. A proton is made of three elementary particles (quarks) and considered to be stable in the standard model, but if a monopole passes nearby, the interaction generates a decay of the proton to other elementary particles. If this decay is observed, it might be the evidence of the existence of solitons.[21] Since all matter have protons and so there are much amount of the protons around us, it is considered that the observation of the rare decay like this might be possible. However, it hasn't been detected so far, in experimental observations. The present situation is that the value of the upper limit of the monopole density in the universe is computed from the data.

As a matter of fact, the monopole density in the universe is associated with so-called "monopole problem", and which is a big problem in cosmology (see Chap. 6.2). It is calculated that, if too many monopoles are created in the early universe, the mass energy of the monopoles hold a big portion of the energy of the whole universe, then the universe doesn't inflate but contract soon so that it does not grow up to become the present vast universe. Hence, a mechanism to decrease the monopole density is required. The most promising candidate of it is "inflationary cosmology". The inflationary cosmology is currently studied by cosmological physicists actively, and an interesting relationship with string theory/D-branes has been revealed. I will give an explanation of this theme in Chap. 6.2.

In addition to the monopoles, vortices also may exist in the universe as well, and might be observed. However the vortices are solitons in 2-dimensional space, while the space of our universe is 3-dimensional. How can we resolve this gap of the dimensions? This problem is related with a very interesting theme of dimensions of solitons, and I will give a full explanation in Sect. 3.1.

2.4.2 Duality and Soliton: Solving Difficulties in Field Theory

Let us explain a theoretical importance of the solitons. This is also the importance of the existence of solitons (= D-branes) in string theory, and the most important point for establishment of the ultimate theory. Solitons have a possibility to resolve a certain kind of theoretical difficulties in field theories. First, I have to explain what the difficulties are. And after that, I will describe the role of the solitons as a method for resolving the difficulties. The method is a very strange symmetry "exchanging elementary particles and solitons"!

---

[21]There are various other physical mechanisms which generate a decay of protons. In reality, we need to examine how the protons decay, for identifying whether it is by monopoles or not.

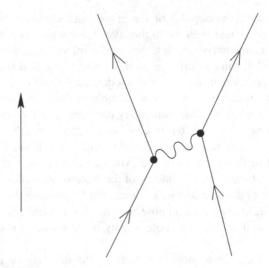

Fig. 2.17 A Feynman graph stands for a scattering process of two electrons. The *solid lines with arrows* stand for the electrons, and the *wavy lines* stand for the gauge field of electromagnetism (photon). The *arrow* on your *left* depicts the direction of time. The horizontal direction shows space

Basically in field theories, once the action is given, we can calculate any physical quantities. Here, I mean by "we can" that we can do it in principle, but to actually perform the calculations exactly is a hard task. There, difficulties in the calculations often appear. Even if the way of calculating physical quantities and the equations are given, it is difficult to actually calculate them and derive the answer.

Let us explain an abstract of how to calculate a collision/scattering process of two electrons in a field theory. In the calculation of the field theory, as a starting point an action written with a field of the electron and the electromagnetic field is used. This is called Quantum Electrodynamics. The scattering of the electrons is given by a graph like the one in Fig. 2.17. This is called a Feynman graph, and it also stands for a way of the calculation. Once a graph is given, we can write a mathematical formula showing the scattering probability corresponding to it. In this graph, the vertical direction shows time and the horizontal direction shows a space. (Though the spacetime is $3 + 1$-dimensional, as a matter of convenience for showing a figure, we depict the space in a 1-dimensional manner.) Lines are "worldlines" of the particles, that is, trajectories of the particles in the spacetime, and the point of intersection of the line with a horizontal line stands for the location of the particles at that time. Solid lines are the electrons and wavy lines are the electromagnetic field. The scattering by an interaction among electrons comes from the interaction by the electromagnetic field, and it correspond to exchanging an elementary particle of the electromagnetic field (a photon). The configuration is shown in the Feynman graph and with that the probability for the scattering of the electrons is calculated. Since the "coupling constant" of the interaction between one electron and the electromagnetic field is the electric charge e, the magnitude of the electron scattering is about e^2 in magnitude (this is called a scattering amplitude, squared of the absolute value of which gives the probability of the scattering). This is because in the Feynman graph of Fig. 2.17 there are two points of the interaction.

In order to calculate the scattering of the electrons in the field theory, at first, while fixing the condition that two electrons come in and two go out, we write down all possible Feynman graphs. In the quantum electrodynamics, in addition to the Feynman graphs of Fig. 2.17, there are innumerable Feynman graphs with higher powers of e (called orders), and the sum of all the terms with the higher orders is the final description of the scattering. For example, there is a graph like Fig. 2.18.[22] This is order e^4. A calculation method like this using series-expansions of graphs is called a "perturbation theory." The perturbation theory is a fundamental and an important method of calculations in field theories.

As a matter of fact, it is quite difficult to calculate terms of higher orders and it is a laborious problem to sum the terms of all the orders. However in this case of QED, since the quantity e^2 called the fine structure constant is given approximately by $1/137$ (in the unit of $h/2\pi = 1$ and $c = 1$ where h is the Planck constant and c is the speed of light), the whole can be approximated by the first term. That is, the higher order terms are given as small perturbations. So, by calculating higher order terms and adding them, the calculation gets more precise. The perturbation theory of QED have achieved a huge success. For instance, experimental results of the magnetic moment of an electron agree with theoretical calculations of QED at the

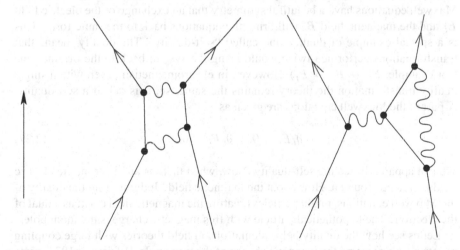

Fig. 2.18 An example of a Feynman graph of higher orders. In the previous graph, there are two interaction points and so it gives e^2 in magnitude, while in this graph it is e^4 in magnitude

[22]When the Feynman graph includes a loop, the result of the calculation is infinite because of a sum of all possible energy of particles rounding along the loop. This infinity can be made finite by a procedure called "renormalization." According to this recovery, J. Schwinger, R. Feynman, and S. Tomonaga awarded a Nobel prize in 1965 for their theoretical contribution to QED.

accuracy of more than ten digits. This can be called the most successful theoretical calculation ever.

The standard model of elementary particles includes not only QED as a part but also quantum chromodynamics (QCD) as other important part. QCD is a field theory describing quarks and interactions between quarks which form protons, neutrons and mesons. Perturbation theories actually make sense only when their coupling constants are small, but unfortunately the coupling constant of QCD is large and of order 1 in magnitude at the energy region of our interest. So, there the perturbation theory doesn't make sense any more. This is a serious difficulty on calculations in field theories and in the standard model of elementary particles.

Then, how can we calculate QCD? An answer is to use computers. Instead of using perturbation theories, actually we can calculate the whole process in computers. By discretizing points in spacetime and making them a lattice, we can calculate the field theory without the perturbation theories on the computers. This is called lattice QCD, and it is one of main subjects in elementary particle physics. Due to the limitation of the computer ability, the lattice size of each side in the spacetime is less than 100. However, the computer results give good achievement.

Then, is it impossible to perform by hand the calculations of theories having large coupling constants such as QCD? To solve this problem, the concept of "duality" using solitons comes up here. The duality means a situation where two different physical systems (or physical descriptions) turn out to be equivalent to each other.

I will explain what this duality is, with an example of electromagnetism. The Maxwell equations have a beautiful symmetry that an exchange of the electric field E_i and the magnetic field B_i will bring the equations back to the same form. This is a special example of duality, and called "self-duality." The duality means that transformations performed twice would bring the system back to the original one (for example, $E_i \to B_i \to E_i$). However, in electromagnetism, even with a single duality transformation the theory remains the same, so, it is called a self-duality. A part of the Maxwell equations are given as

$$\partial_i E_i = 0, \quad \partial_i B_i = 0. \tag{2.27}$$

We can apparently see the self-duality. Now, when there is an electron, the electric field is created around it. How about the magnetic field? Judging from the duality of the Maxwell equations, might particles creating the magnetic field exist, as a dual of the electron? The hypothetical particle with this magnetic charge is the monopole.

Let us see how the duality help calculations in field theories with large coupling constants. Concerning the magnetic charge of the monopole, P. Dirac in 1931 wrote down the following constraint equation: $g = 1/2e$. This constraint equation is derived from the condition that electrons behave in a consistent manner around the monopole quantum-mechanically.[23]

---

[23]Let me explain how to get the Dirac's condition. In quantum mechanics any particle is represented by a description of waves, a wave function $\psi(x)$. The square of the absolute value of

We have an interesting finding if we combine this condition with the electromagnetic duality. When the electric field is exchanged with the magnetic field the electrons are exchanged with the monopoles, e and g should be exchanged with each other. According to the Dirac's condition, g is large for small e. In this sense, the theory with weak coupling ($e \ll 1$) is equivalent to a theory with strong coupling ($e \gg 1$). Even when we cannot apparently use the perturbation theory because of the large coupling constant e of the theory, if there is the duality, the theory is equivalent to a theory with weak coupling constant, so we can do calculations using a perturbation theory.

However, so far, any monopole of this electromagnetism has not been found yet. There is a theoretical reason about it: It can be shown that the magnetic field $B_i(x)$ around a point-like source of it cannot be written by any gauge field. $A_\mu(x)$.[24] However, not in the case of the monopoles of this electromagnetism, but of monopoles as solitons in the non-Abelian gauge theories mentioned earlier, there is no such difficulty, and we can find a soliton as a field configuration with smooth, finite and localized energy. Here, the fact that electromagnetism is generalized to be a non-Abelian theory is helpful. We can actually expect the theory like this has a self-duality. However proving it mathematically is very difficult. This is because, of course we cannot use the perturbation theory for the proof. And this is also because we exchange elementary particles and solitons. Theoretical properties beyond the perturbation theory such as the dualities is called non-perturbative. It is the very important to understand physics theory in an essential way.

Though Maxwell electromagnetism is self-dual, there are cases which are not self-dual. A duality transformation brings a certain theory to a completely different theory, while they are equivalent to each other. In most cases, on the one hand the coupling constant is small while on the other it is large. And elementary particles on the one hand are interpreted as soliton on the other. This is a duality among different theories. In this kind of duality between strong and weak couplings (strong–weak duality), in the strong coupling regime of a certain theory another equivalent field theory plays a vivid role, and there solitons of the former theory become a fundamental field. That is, solitons are quite important objects when we want to handle the non-perturbative region of field theories.

---

the wave function $|\psi(x)|^2$ stands for the probability for observing the particle at the point x. Now when the monopole is placed at the origin $x = 0$, suppose that electron is put near the monopole and then let the electron go round about the monopole. When it comes back to the original place, the wave function of the electron must be equal to the original wave function. In the process, a calculation shows that the wave function acquires a phase factor $e^{4\pi i e g}$ when the electron goes round. The condition of this phase being equal to 1 is the Dirac's condition mentioned above.

[24]Let me quote a proof of it for readers who are interested in it. When there are electric charges like electrons, the right hand side of the first equation of the Maxwell equations (2.27) is not set to 0, but there a density function of the electrons appear. Therefore if there is a magnetic charge, in the second equation of the Maxwell equations (2.27), the right hand side is not 0. But by using the definition equation $B_i = \frac{1}{2}\epsilon^{ijk} F_{jk}$ of the magnetic field and the definition equation (2.1) $F_{\mu\nu}(x) = \partial_\mu A_\nu(x) - \partial_\nu A_\mu(x)$ of the gauge field strength, this is shown to be impossible.

The duality plays a very important role in string theory. The duality makes clear the relation between different string theories. And what is better, thanks to duality, we can handle physics at the region of a strong coupling constant in string theory. We shall describe the duality of string theories in Sect. 4.2, where we find also that solitons of string theory are D-branes, so considering all of them, we see that there exists a symmetry of exchanging strings and D-branes! Therefore, even though we start thinking of string as a fundamental matter in string theory, it turns out that we might actually regard D-brane as fundamental. This surprising result is brought by the duality. I will have a full account of this story in Chap. 7. What is the fundamental ultimate theory describing the whole physics in our universe written by? This is the a very important question. It might be the D-branes, the extended objects, rather than the strings!

And the fact that the D-branes appear concerning the duality in string theory gives a big influence on field theories of elementary particles as well as string theory. Let us get back to the previous problem of QCD. It is not known whether the actual QCD in the standard model of elementary particles has such an exact dual theory as I mentioned here. However, by recent advances in string theory, we start to learn that a certain gravity theory describes the strong coupling region of theories similar to the QCD. This is a great progress in theories. First of all, what is strange is that QCD, the theory of quarks, can be describe by a gravity theory! What is more, this duality requires the gravity theory to be not in $3 + 1$ dimensional spacetime but in $4 + 1$ dimensional spacetime! This is a duality of theories in different spacetime dimensions, and called holography. ("Holography" originally means a principle of 3-dimensional photograph (hologram), while in physics it is used for the meaning I mentioned.) If this new duality is developed more and more, analytical calculations by hand can attain various calculations of QCD, and the day when we can achieve it might be a near future. D-brane makes clear the following points: why the theories with different dimensions are equivalent and why a gravity theory and a non-Abelian gauge theory are equivalent. I will have a full explanation of the mechanism in Chap. 6.4.

Chapter 3
Dimensions of Solitons, Dimensions of String Theory

In Sect. 2.4, I focused on the importance of solitons. The importance of course depends on what kind of field theory and solitons we consider. Solitons appear in various field theories, and an index characterizing those various solitons is the dimension of the solitons. The kinks, vortices, and monopoles having appeared in the preceding chapter arc classified by their dimensions.

In this chapter, at first, we consider various dimensions of field theories and solitons, in Sect. 3.1. We will see an interesting fact that physical applications of identical solitons are completely different according with which dimension in spacetime they are put. This more often occurs in string theory. This is because string theory is basically a theory in higher dimensional spacetime, as will be explained in Sect. 3.2. Solitons appearing in string theory (namely D-branes, which are the main subject of this book) have a vast variety, and their applications for physics appear very differently according with their dimensions and their kinds. We will enjoy the variety and the attraction of the applications in Chap. 6. In Sect. 3.1, first let us see how solitons described so hitherto change their appearance, according to the dimensions. If we consider solitons in the situation of spatial dimensions larger than three, over stretching our imagination, a quite attractive idea called "braneworld" comes out. There, an interesting idea of "people living on the soliton" emerges. We might live on the soliton, in reality.

Solitons in higher dimensions are a basis of the concept of D-branes. I will introduce the D-branes as higher dimensional solitons in string theory in Chap. 4, and in Sect. 3.2 I will explain a basic part of string theory as a necessary preparation for the introduction of D-branes. String theory brings us extremely exciting physics, even without mentioning D-branes. I will have an explanation on, how elementary particles in the world can be expressed by strings and what the higher dimensional spacetime appearing in string theory is, and how string theory is related consistently to the fact that dimensions of our space is three.

K. Hashimoto, *D-Brane*, DOI 10.1007/978-3-642-23574-0_3,
© Springer-Verlag Berlin Heidelberg 2012

3.1 Dimensions of Solitons and Braneworld

3.1.1 Solitons and Dimensions

Since the dimension of the spacetime we live in is 4 consisting of one-dimensional time and 3-dimensional space (this is often written as 3+1 dimensions), realistic field theories are written by function fields of 4-dimensional spacetime coordinates x^μ ($\mu = 0, 1, 2, 3$). For example, electromagnetism and the standard model of elementary particles are like that. However in even lower dimensions, realistic field theories (namely which can be compared with experiments) exist. For instance, the physics of conductors on spatial 2-dimensions (such as a board) or 1-dimension (such as a line) is one of the cases. Or even for the case of dealing with elementary particles, when we treat not many elementary particles at once in a field theory but only a single particle (namely the case of one body problem), the position of the elementary particle is given by a function $X^i(t)$ ($i = 1, 2, 3$), therefore, this is regarded as a field theory in 0+1 dimension. In this sense, particle mechanics can be regarded as a field theory in lower dimensions. In this way, in physics, there appear field theories in various dimensions in accordance with physical systems we consider.

In Sect. 2.3, we saw three kinds of solitons relating with their dimensions and the symmetries. The first soliton is the kink soliton which is static in one-dimensional space (1+1 dimensional spacetime), the second one is the vortex in 2+1 dimensional spacetime and the third one is the monopole in 3+1 dimensional spacetime. Which, among these three, can exist in the 3+1 dimensional spacetime which is familiar to us? The answer is, in fact, not only the monopole, but all of the three.

Let me explain the reason. We shall remind that in the example of the first soliton, a water wave in Sect. 2.2, the energy is localized about the direction x, while on the other hand it is not localized about the direction y, and what is more, it is completely independent of the coordinate y. As is obvious from this example, a water wave in the 1+1 dimensions (the time t and the space x) may be regarded also as a soliton in 2+1 dimensions (the time t and the space x, y). Though the whole energy of the soliton in 2+1 dimensions is infinite (because of the integration about the direction y), this is not a problem. The physically important quantity is the energy of the soliton in the 1+1 dimensions, and in terms of the 2+1 dimensions, it is the energy density per a unit length of the direction y.

From the example of this water wave, it is easy to learn that even the kink or the vortex could become a soliton in 3+1 dimensional spacetime. Let us see Fig. 3.1. If the kink or the vortex exist in 3-dimensional space, the kink extends in a 2-dimensional space, or the vortex extends in a 1-dimensional space. Such an extension of the solitons put in a higher dimensions is called "worldvolume." (See also Fig. 4.1.) The dimension of solitons is specified by the spacetime dimension of the physical theory we consider and the spacetime dimension of the worldvolume. In the case of this kink in the 3+1-dimensional spacetime, its worldvolume has 2+1 dimensions. On the other hand, in the case of the vortex in the 3+1-dimensional

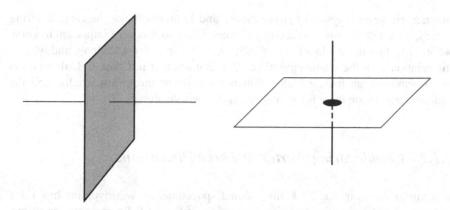

Fig. 3.1 *Left*: The figure of a kink in a 3-dimensional space. The horizontal axis is the original x^1 direction considered in the ϕ^4 model, and the energy is localized at a point on that axis. If the space spreads into a 3 dimensional space, the lump of the energy homogeneously distributes along the expanded directions (the vertical direction and the forward direction), and it has an extended worldvolume. That is, the kink is a wall. *Right*: The figure of a vortex in a 3-dimensional space. The horizontal directions are x^1 and x^2 at which the vortex is originally localized. The vertical direction is a newly added dimension, and the vortex extends along that direction. Namely, the vortex becomes a vortex string

spacetime, the worldvolume has $1 + 1$ dimensions. And finally in the case of the monopole, the worldvolume is of 0+1-dimension. In particular, $1 + 1(= 2)$-dimensional worldvolume is called a worldsheet, and that of $0 + 1(= 1)$ dimension is called worldline. When defining the dimensions of the worldvolume, notice that we ignore the thickness of the soliton and count it as infinitely thin.

It is obvious that the way of thinking of the worldvolume of solitons is appropriate not only for the case of the 3+1 dimensions in spacetime for the field theories of concern. For example, let us consider a kink solution in D-dimensional spacetime. The worldvolume of the kink solution must have $D - 1$ dimensions. Or for the case of a vortex in the D-dimensional spacetime, the worldvolume has $D - 2$ dimensions. In this way, solitons in field theories are characterized not with the spacetime dimensions of the field theory, but with how many dimensions the solitons worldvolume is smaller by compared to the spacetime dimension of the whole field theory. This is called co-dimension. The kink has one co-dimension, and the vortex has two co-dimensions, and the monopole has three. By using this co-dimension, we can draw out the characteristic of solitons in field theories in any dimensions. The worldvolume is a useful notion to designate the dimension of an extended object floating in a certain dimensional space time, not only for solitons. For example, in string theory, an object extending along a 1-dimensional space in D-dimensional spacetime is fundamental. For this case, the worldvolume is $(1 + 1 =)$2-dimensional, and so is a worldsheet. Let us remember this notion of the worldvolume, as we will often use it when D-branes of various dimensions appear later.

In this way, we begin with considering various spacetime dimensions and worldvolumes. You might think that we define somewhat easy stuffs in a difficult manner. However, physics of string theory and D-branes is not the case. In string theory, higher dimensional spacetime is considered, so it is very important to know which direction the worldvolume of solitons and strings extends along, and what is the relation with the whole spacetime. You can see that in Chap. 6. Only masters of dimensions can handle higher dimensional spacetime unrestrictedly, and the fundamental notion there is the concept of this worldvolume.

3.1.2 Cosmic String: Vortex in Three Dimensions

If a vortex exists in our $3+1$ dimensional spacetime, its worldvolume has $1+1$ dimensions. Therefore it is a string-like object in the $3+1$ dimensional spacetime. This is called a vortex string. It is a very important object in various physics. The most famous example is a state in which a magnetic field is confined and forms a vortex string, in a superconducting matter. Here, I will describe an example related with elementary particle physics and cosmology, for our later convenience.

The monopoles are a soliton appearing under the symmetry breaking in a grand unified theory, and a vortex string can also be generated at this symmetry braking. This is called a "cosmic string." The cosmic string is an extremely long soliton which extends over the whole universe and so it is an interesting object, while it has not been observed yet.[1]

If there is a string, it bends the trajectory of the light traveling around it, because of the gravity generated by the mass energy. This is one of the phenomena called "gravitational lensing effect," and galaxies and such at the other side of the cosmic string seen from the earth can look doubled. We can judge whether the gravitational lensing effect comes from the cosmic string or any other localized celestial body, by analyzing the pattern of the duplication. How interesting if we could observe magnificently huge solitons lying in the universe!

In addition, interestingly, these days a possibility was pointed out that this cosmic string may be a fundamental string or a D-brane in superstring theory. The fundamental matters, strings or D-branes, might float in the night sky.

By the way, although too large number density of monopoles is a cosmological problem, the cosmic strings escape this problem very well. Suppose that there are a lot of strings in a 3-dimensional space. They move and inevitably collide each other, then reconnections take place (see Fig. 3.2). The reconnection makes the string to form a closed loop, and it disappears by shrinking the radius of the loop. Namely, cosmic strings have a self-annihilation mechanism, as opposed to the case of the monopoles. The mechanism of the reconnection of the cosmic strings is an

---

[1]Though in the 1990s cosmic strings were considered to play an important role in density fluctuation structure formation of the universe, recently it is being denied by a precise observation of cosmological background radiation at 3 Kelvin. However, it is still possible that the cosmic string exists in the universe.

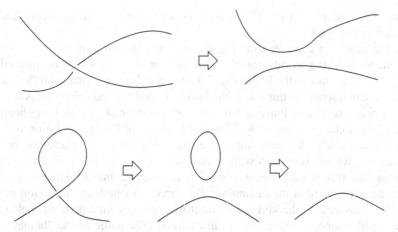

Fig. 3.2 An example of a reconnection process of cosmic strings. In the *lower figure*, by the reconnection a loop is formed, and it shrinks and finally disappears

interesting subject of physics in cosmology. The mechanism differs for strings of string theory, or the vortex solitons, or D-branes. This difference comes up as a difference of the probability of the reconnection at the collision, and it is used to find the identity of the string, at the step of observing actually something like cosmic strings. Let us see it concretely in Sect. 5.3.

3.1.3 Braneworld

In the case of the cosmic string, we considered the situation that the vortex itself extends spatially in one dimension, by upgrading the vortex to an object in a $3+1$ dimensional spacetime. Generalizing this method further, let us imagine the case that the worldvolume of a soliton is our $3+1$ dimensional spacetime. This is the very beginning of the attractive idea "brane world."

Let us consider a kink, the simplest soliton. When we introduced the kink in Sect. 2.3, it is a localized object in one spatial dimension, and the dimension of the worldvolume is only one, the time. However, this story can be easily extended to general dimensions. Consider the ϕ^4 model in D dimensions. The spacetime is spanned by $x^0, x^1, \cdots, x^{D-1}$ and the field ϕ is a function of these. In order to solve its equation of motion, for simplicity we suppose that the field ϕ doesn't depend on $x^2, x^3, \cdots, x^{D-1}$. Then, the static equation of motion is the same as the previous (2.16), and as a result of this, the same solution (2.17) exists. This is an object of co-dimension 1, and the dimension of the worldvolume is $D - 1$. To reduce the field at higher dimensions to a lower dimensions, by assuming that the field is independent of the coordinates, like this, is called a dimensional reduction from a higher dimensions. In the present case, concretely, we consider only a field which is independent of $x^2, \cdots, x^{D-1}$. This method is often used in string theory, and so used also in this book.

By the way, as a concrete example, if we start with our familiar $D = 3 + 1$, the dimension of the worldvolume of this soliton is 2+1, which is spanned by x^0, x^2, x^3. As we saw in the left of Fig. 3.1, this is a 2-dimensional wall floating in a 3-dimensional space. In this sense, the kink is called also domain wall. And, now if we increase the whole dimension by one and consider $D = 4 + 1$, applications of the physics change completely. The worldvolume of the kink solution is 3+1-dimensional, namely, the dimension of the spacetime we live in. Therefore, we can consider that we live only on a soliton and the actual spacetime is 5-dimensional. The idea like this is called braneworld.[2] It is essential that a high dimensional spacetime is prepared at the beginning. The direction (the fifth dimension in this example) transverse to the kink (our spacetime), namely directions not belonging to the worldvolume, is called "extra dimensions." The name of the "braneworld" comes form the D-branes. The braneworld needs a higher dimensional spacetime, while string theory needs inevitably a higher dimensional spacetime and solitons appearing there are the D-branes. The braneworld is the idea which have been developed there. The appearance of the higher dimensions in string theory will be explained in the next section, and the D-branes in Chaps. 4 and 5, and after that, I will have a full account of this fascinating braneworld in Chap. 6. However, before that, in order to make sense of the braneworld, I must define physically the sentence "to live on a soliton." Let us think of this next.

3.1.4   How to Live on Solitons?

If the braneworld is real, why don't we feel the higher dimensional space even though it is there? To tell the truth, on the worldvolume of the soliton, there exists a field theory standing for a motion of the soliton and such, and it composes the physics of people living on the soliton. Here, let us construct the field theory only on the worldvolume, concretely for the kink solution (2.17).

By using the dimensional reduction, let us think of a kink in a D-dimensional spacetime. The worldvolume of the kink is $(D-1)$-dimensional spacetime spanned by $(x^0, x^2, x^3, \cdots, x^{D-1})$. Though the solution is the same as the previous (2.17), we generalize this ϕ and suppose that X^1 is not constant but a function of $x^0, x^2, x^3, \cdots, x^{D-1}$. (Here, note that X^1 does not depend on x^1.) Then, of course, (2.17) is not in general a classical solution, but this turn out to be again a solution only when $X^1(x^0, x^2, x^3, \cdots, x^{D-1})$ satisfies a certain equation. The equation is given by

$$\left[-(\partial_0)^2 + (\partial_2)^2 + (\partial_3)^2 + \cdots + (\partial_{D-1})^2\right] X^1 = 0 \qquad (3.1)$$

---

[2]This idea was proposed by K. Akama (1982), V. A. Rubakov and M. E. Shaposhnikov (1983).

and this is the same as the equation of motion of a massless field "living" on the $(D-1)$-dimensional spacetime. Here, we approximate X^1 to be small enough, when deriving (3.1). Since this equation for the new field X^1 is the same as the one derived from an action in the $(D-1)$-dimensional spacetime,

$$\int dx^0 dx^2 dx^3 \cdots dx^{D-1} \left[-(\partial_0 X^1)^2 + (\partial_2 X^1)^2 + \cdots + (\partial_{D-1} X^1)^2 \right],$$

we can say that X^1 is subject to this action. Namely, X^1 lives on the $(D-1)$-dimensional spacetime which is lower by one dimension compared with the original D dimensions. The X^1 field theory defined by this action is called an effective field theory of the kink. The physical meaning of the field $X^1(x^0, x^2, \cdots)$ is obvious. The original worldvolume of the kink solution is spanned by $x^0, x^2, \cdots, x^{D-1}$, and X^1 is given as a function of those, and the kink position in the direction x^1 is provided by the value of X^1 (see Fig. 3.3). In other words, the shape in the higher dimension of the kink is given by the field function X^1.

Since the field X^1 does not depend on x^1 and is a field in $(D-1)$-dimensional spacetime, it is considered to live only on the kink. Let us take a look. If we expand the solution (2.17) by X^1 by supposing that X^1 is small, we obtain

$$\phi = \hat{\phi}(x_1) \Big|_{X^1=0} - X_1(x^0, x^2, \cdots) \left[\partial_1 \hat{\phi}(x^1) \right]_{X^1=0} + \cdots . \tag{3.2}$$

This first term is the original solution (2.17), and the second term is the term which shows how much the field X^1 of our concern extends along the actual direction x^1. That is, we regard the soliton (the first term) as a "background" and an infinitesimal difference from that is the field appearing in the field theory. This second term is proportional to $\left[\partial_1 \hat{\phi}(x^1) \right]_{X^1=0}$, and as you see from the original shape of the solution (2.17), it is localized around $x^1 = 0$ (see Fig. 3.4). Namely, the field X^1 lives only on the soliton. In this way, it is shown that a massless field X lives on

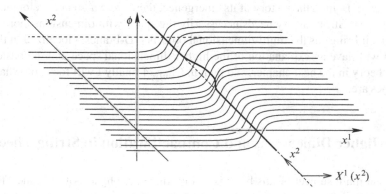

Fig. 3.3 The center point of a kink solution depends on x^2. The center point is denoted by a *bent bold line*. This center position is given as a function $X^1(x^2)$

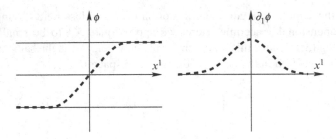

Fig. 3.4 *Left*: The kink solution with $X^1 = 0$. *Right*: a derivative of the kink solution by x^1. This means an extension of the field X^1 into the higher dimensional direction. We can see that the field X^1 is localized at the position of the kink

the worldvolume of the kink solution, but actually this fact is quite general. For example, for a vortex or a monopole, on its worldvolume it is known that massless fields corresponding to the position of the soliton exist, as many as the number of the co-dimension. That is, two (three) massless fields appear on the worldvolume of a vortex (monopole). The number of the field appearing can be understood by reminding that the parameter showing the position of the kink is upgraded to the massless field X. This emergence of the massless fields is comprehended in terms of theoretical physics, as in the following. There is a translational symmetry in the whole spacetime when the soliton does not exist. On the other hand, the translational symmetry along the co-dimension direction is broken if the soliton exists. Applying with a theorem (Nambu–Goldstone theorem) that any symmetry breaking accompanies a massless particle, we can understand that the massless field X appears corresponding to the direction transverse to the worldvolume of the soliton. In this way, the braneworld is introduced as a soliton in a higher dimensional spacetime. The new physics brought by the existence of the extra dimensions is too extensive to introduce here, and it predicts novel experimental consequences. I will explain a part of them in Chap. 6. It may seem that the braneworld is a notion which have necessarily appeared, as you follow this book until here. However, it is actually not the case. Behind the history of its emergence, there were a great development in string theory. String theory needs necessarily the space with dimensions more than four, and it brings us the multi-dimensional objects – D-branes – localized in there. First, I will have an explanation of the higher-dimensional space and the basics of string theory in the next chapter, and then in Chap. 4 finally I will focus on what the D-branes are.

3.2   Higher Dimensions and Compactification in String Theory

While the main subject of this book is to introduce the physics of D-branes, first I have to describe the spacetime dimension of string theory itself which is a stage the D-branes of various dimensions play a vivid role on. Let us see that in string theory higher-dimensional spacetime inevitably appears and it is the origin that makes physics of string theory rich and fruitful. String theory is a theory in which point-like particles that are fundamental constituent elements in the theory of particles are replaced by strings extending spatially in one dimension. This replacement changes dramatically the theory. One of the remarkable differences is that the dimensions of the spacetime in which strings exist turn out to be determined by a consistency of the theory. (In the case of the theory of particles, the dimensions of the spacetime in which the particles move is basically not constrained.[3]) This is a wonderful property of string theory. The spacetime in which strings move is called "target spacetime." The reason of the limitation of the dimensions of the target spacetime comes from a natural requirement that there should exist Lorentz symmetry there.[4]

The dimensions of the target spacetime must be 26 or 10. This 26 is for the case of bosonic string theory, on the other hand the number 10 is for superstring theory. (I will explain these species of string theories soon later.) Therefore, if we consider string theory, we are inevitably thrown into the higher-dimensional spacetime. If the higher dimensions like these appear, you might think it is not useful as a realistic physical theory any more. As a matter of fact, historically, string theory appeared as a theory describing particles called hadrons such as protons. However, because of this limitation on the dimensions, it lost its status as a theory of hadrons. However, in late 1980s, thanks to the developments on the method "compactification" which I introduce later, a way to "compactify" the ten dimension to the 4-dimensional spacetime has been constructed, and string theory started to be put in a spotlight as a unified theory or an ultimate theory.

3.2.1   Species of String Theories

First, let me briefly explain the bosonic string theory and the superstring theory which are the basis of the stories on D-branes. For that, it is easier to focus at first on what the theory of a point particle is. Since the worldline trajectory of the point particle in spacetime extends in one dimension, let us write the coordinate parameterizing it as τ. Then, the worldline is given as a function $X^\mu(\tau)$. This $X^\mu(\tau)$ is a function specifying a point X^μ in the target spacetime once a certain point τ on the worldline is given. If the target spacetime is D-dimensional, μ, the index of the coordinates of the target spacetime, runs from 0 to $D - 1$. If we

---

[3]The spacetime dimensions are restricted in order to renormalize infinities in calculations of Feynman diagrams, but the constraint is not so strict as string theory. For example, even in 4-dimensional spacetime, one can write infinite kinds of field theories for particles, but we have limited kinds of string theories.

[4]Lorentz symmetry is a symmetry required by the special relativity. It is a generalization of a rotational symmetry of space to include also the time direction. Even for theories of particles, it is definitely required when we consider relativistic particles. The fact that the requirement of the Lorentz symmetry determine the dimensions of the spacetime in string theory is a result of a quantum mechanical treatment of string theory, and here I will not describe the mechanism.

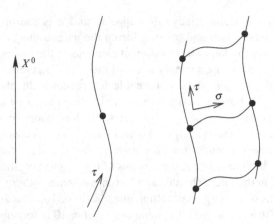

Fig. 3.5 A worldline of a point particle and a worldsheet of a string (*Right*). The *left arrow* stands for the time direction X^0 of the target spacetime

use the simplest parameterization $X^0 = \tau$, $X^i(X^0)$ stands for the spatial position ($i = 1, 2, \cdots, D-1$) of the particle at a certain time X^0. You may be familiar with this way of writing in mechanics.

Now, in the case of string theory, this worldline is replaced by a worldsheet, which is parameterized by two coordinates (τ, σ) (see Fig. 3.5). That is, the position of the string in the target spacetime is provided by a function $X^\mu(\tau, \sigma)$. The theory described by this $X^\mu(\tau, \sigma)$ is called bosonic string theory. This might be regarded as a field theory in $1 + 1$ dimensions. This is because the function field X^μ is written by the two coordinates τ and σ. The reason why it is called bosonic is that the function field $X^\mu(\tau, \sigma)$ is a bosonic field in two dimensions. A "boson" means a field with an integer spin (spin is an inner angular momentum intrinsic to elementary particles), in general. (On the other hand, fields with half-odd integer spins are called "fermions." For example, in the standard model of elementary particles in $3 + 1$ dimensions, electrons are fermions and the gauge fields $A_\mu(x)$ of electromagnetism are bosons.) $\mu = 0, 1, \cdots, D-1$ is an index standing for directions of the target spacetime, but as seen from the viewpoint of the field theory in $1 + 1$ dimensions, it is just an index labeling the bosonic fields $X(\tau, \sigma)$. Furthermore, as mentioned earlier, this string theory turns out to have $D = 26$. Strings have lengths, which is represented by a period $0 \leq \sigma \leq 2\pi$ of the coordinate on the world sheet. In the case of a closed string forming a loop, we impose the following periodic boundary condition,

$$X^\mu(\tau, \sigma + 2\pi) = X^\mu(\tau, \sigma) \tag{3.3}$$

and this theory is called a closed string theory. In the case of an open string, we impose a free boundary condition at the end points $\sigma = 0, 2\pi$ of the string, as

$$\partial_\sigma X^\mu(\tau, \sigma) \Big|_{\sigma=0, 2\pi} = 0 \tag{3.4}$$

and this theory is called an open string theory. Here, instead, one can impose a fixed boundary condition. it actually is related to the definition of the D-branes which appear in Sect. 4.2. The "D" of the D-branes is an abbreviation of the fixed boundary condition, namely, the Dirichlet boundary condition. Here let us consider only the free boundary condition (called the Neumann type), first. The superstring theory is a theory contains the supersymmetry as it is considered as a field theory on a worldsheet (that is, a $1 + 1$-dimensional spacetime). The supersymmetry is a symmetry which interchanges bosons and fermions, and it has a mathematically beautiful structure. Since this beautiful structure brings us physically important characteristics, physicists respect theories possessing the supersymmetry in elementary particle physics. String theory has also the tendency, and it is often told that "Superstring theory doesn't contain tachyons and thus is consistent." Later, I will have an explanation of the spectrum of string theory, with regard to it. Because of the supersymmetry on the worldsheet, superstring theory contains also fermion fields on the worldsheet of $1 + 1$ dimensions, in addition to the bosonic fields $X^\mu(\tau, \sigma)$. Here, we omit the details as they are unnecessary.[5] By the influence of existence of these fermionic fields, the dimensions of the target spacetime of the superstring theory is determined to be 10. As a result, the index of X^μ runs as $\mu = 0, 1, \cdots, 9$. As briefly summarized, there are two kinds of strings, open and closed, and for each, there are two kinds, bosonic strings and superstrings. Combinations of these form the whole species of superstring theories. We must pay attention to the fact that there is no theory of only the open strings. As you can see the worldsheet of the left figure in Fig. 3.6, the end point of an open string can joint to become a closed string. Or, depending on how you cut the worldsheet, you can find a closed string in an open string worldsheet (Fig. 3.6, Right). Therefore, for the open string theory itself to be consistent, it must contain closed strings. On the other hand, in the case of the closed string theory, there is a requirement that its worldvolume has no boundary from the first place, there is no inconsistency in a theory only with closed strings. However, if one considers D-branes which are solitons in string theory, one can see a surprising requirement that the closed string theory must contain open strings too. I will explain this fact in Sect. 4.2.

3.2.2 Spectrum of String Theory: Relation Between Particles and Strings

In the bosonic string theory or in the super string theory, we start with a certain kind of a string. Then, how can it describe all the particles and interactions in our world

---

[5] Strings oscillate in spacetime. In the case of the closed string theory, the oscillations propagating on the closed sting are either right-moving or left-moving. Now, there is a hetero-type string theory in which the left-movers are bosonic while the right-movers are of a superstring. This is called a heterotic superstring theory. Since it is useful to construct grand unified theories of elementary particles from the string theory, the heterotic string theory is enthusiastically studied as a candidate of the unified theory.

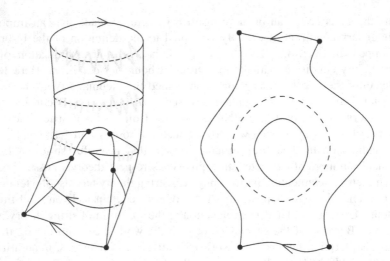

Fig. 3.6 Worldsheets of a string. The *arrow* stands for the direction of the coordinate σ on the string. *Left*: The end points (shown as *black blobs*) of an open string joints to form a closed string. *Right*: a worldsheet in which an open string make a loop is shown, but if we look at a part of this worldsheet as the *dashed line*, it turns out that a closed string appears. Therefore, in the theory of open strings, closed strings should be contained

in a unified way? In order to explain this, we have to know the "spectrum of string theory" and the "compactification." At first, I will have an explanation of the former "spectrum of string theory." Interestingly, in string theory, if one considers a single string, one can express infinite kinds of particles by the oscillations. Let us show this in a simplified example. We can regard an open string with the free boundary condition as a water surface in a swimming pool (with a very narrow width so that the surface is thought of as one-dimensional). A position in the swimming pool is specified by the coordinate σ. Standing waves on this water surface can be classified by the number of nodes on the wave (Fig. 1.2). Speaking a little more precisely, any standing waves can be described by a superposition of basic standing waves classified by the number of nodes. That is, the number of nodes can classify the σ dependence of X^{μ}. This can be considered as a Fourier transformation for the direction σ. If the basic standing wave classify and definite the direction σ, the remaining of the worldsheet is only the direction of time τ and it is as same as the point-like particles. As you can understand from this, the single string can represent infinite kinds of particles, which are labeled by the number of nodes of the oscillation on the string. Let us see how these particle states are labeled. Let us imagine that, by quantum mechanics, in all those basic standing waves on the string the magnitude of each is determined by a positive integer number. If you are familiar with quantum mechanics, please read the footnote.[6] Then, the state of the string must

---

[6]The physics of each basic standing wave is equivalent to the quantum mechanics of a harmonic oscillator. For the harmonic oscillator, a quantum state is given by multiplying operators creating waves, on a vacuum state. In the preset case, since the number of species of the standing waves is

be specified by a certain array of non-negative integers: $(N_1, N_2, N_3, \cdots)$. Here, N_n $(n = 1, 2, 3, \cdots)$ shows the magnitude of the standing wave with n nodes, that is, the magnitude of the Fourier amplitude. Once the integer array is given, the state of the string along the σ direction is determined correspondingly, and the particle state represented by the string is specified. The mass of each particle represented by the string is determined by how much the string oscillates. The reason is as follows. A large oscillation would increase the energy of the string, but in the interpretation as a particle, the energy is just some internal energy, so one can interpret that the mass of the particle increases by that amount. Though I omit the derivation, by using a Hamiltonian of 1+1 dimensions for the string, the mass formula of the particle excitations is given as[7]

$$m^2 (= p^\mu p_\mu) = \frac{1}{l_s^2} \left(-1 + \sum_{n>0} n N_n \right). \tag{3.7}$$

l_s appearing here is the unique quantity which has a mass-dimension in string theory, and called the "string length." The physical meaning of it is that $1/(2\pi l_s^2)$ is the string tension. As the tension is larger, one needs more energy to oscillate the string. It is reflected in this mass formula. The distribution of masses of particles appearing in a theory is called a spectrum, so this is the spectrum of the bosonic string theory. The first part of the spectrum looks

$$N_n = 0 : \text{tachyon}, \quad m^2 = -1/l_s^2$$

$$\left\{ \begin{array}{l} N_1 = 1, \\ N_2 = N_3 = \cdots = 0 \end{array} \right\} : \text{massless particle}, \quad m^2 = 0.$$

---

as many as that of nodes, which is infinity, a string is the same as a system with infinite kinds of harmonic oscillators. Let us write the creation operator of the waves with the node number n as α^μ_{-n}. This creation / annihilation operator is given precisely as

$$X^\mu = x^\mu + l_s^2 k^\mu \tau + i l_s \sum_{n \in \mathbf{Z}, \neq 0} \frac{1}{n} \alpha^\mu_n e^{-in\tau} \cos n\sigma, \tag{3.5}$$

where α^μ_n is a Fourier coefficients of the solution of the equation of motion of the bosonic field X^μ on the worldsheet, $\partial_\alpha \partial^\alpha X^\mu(\tau, \sigma) = 0$ with the free boundary condition. The generators have the index of the spacetime μ, because the position operator X^μ has the index of each coordinate of the spacetime, as in the quantum mechanics. Then, a general state of the string is written as

$$|0; k^\mu\rangle, \quad \alpha^\mu_{-1}|0; k^\mu\rangle, \quad \cdots, \alpha^\mu_{-1}\alpha^\nu_{-1}\alpha^\rho_{-3}|0; k^\mu\rangle, \quad \cdots \tag{3.6}$$

Here k^μ is the center-of-mass momentum of the whole string. Although some creation operators α^μ_{-n} are there, those degrees of freedom can be seen as internal degrees of freedom, since they are all on the waves on the string. Then, one can regard this state (3.6) as a particle carrying the momentum k^μ. In this way, string theory can accommodate infinite kinds of particles at once, by a single string.

[7]Quantum mechanically, N_n counts the number of the creation operators α_{-n} for each node number n contained in a state (3.6) of the string, and -1 originates from the zero-point energy in quantum mechanics.

We notice here that, in this bosonic string theory, there exists a state of a tachyon particle for which the mass squared is negative. In the researches in 1970s and 1980s, by this reason the bosonic string theory was considered to be inconsistent, and so it was discarded away from the subject of researches. However, after the active researches in the late 1990s about the tachyonic state, it turned out that this is not an inconsistency but a quite important state concerning a creation and annihilation of D-branes. The wise readers remember that, in Sect. 2.3, tachyons are related with a creation of solitons. Since this relates to the main subject of this book, I will have a full account of it in Sect. 5.3. Before the D-branes appeared, string theories without tachyons were considered to be important. And so the main streams of research moved from the bosonic string theory to superstring theories. I omit here the explanation of the spectrum in the superstring theories, for simplicity. But you may think of it just the same as the bosonic string theory except for the absence of the tachyon part. Then, the important state is the massless state. The massless state is a state with $N = 1$ in the bosonic string theory, that is, with one unit of the excitation of a single standing wave with one node. This standing wave originally come from the field $X^\mu(\tau, \sigma)$ of 1+1 dimensions, which has the index μ. This is the degree of freedom for the direction along which the standing wave points in the target spacetime. According to this, the massless state must have one index μ. Therefore, the massless state is a vector particle in the 10-dimensional spacetime (26-dimensional one for the case of the bosonic string). Massless vector particles are gauge fields. Gauge fields appears along with gauge symmetries, and for example, as we saw in Sect. 2.2, the electromagnetic field is written by a gauge field standing for photons. Namely, a theory of a gauge field appears from the open string theory.

Then, how about the closed string? In the case of the closed string, oscillation modes can be classified by waves right-moving and left-moving on the closed string, and furthermore each wave can be classified by the number of nodes. So, compared with the open string, we have doubled species of the integers N_n labeling the waves. Let us write this as $(N_n, \widetilde{N}_n)$. From a mass formula, in addition to the tachyon particle state as before, a state with $N_1 = \widetilde{N}_1 = 1$ appears as a massless state. This has, as considered in the same way as the open string case, two indices μ of the target spacetime: that is, the index is (μ, ν). Such a massless field should be a gravity field. From the closed string theory, a gravity theory comes out. To be a little more precisely, the part with a symmetrized indices μ and ν after the part proportional to the unit matrix is removed, is the particle of the gravity field (graviton). The anti-symmetric part is a gauge particle called Kalb–Ramond field. The part proportional to the unit matrix has no index and called a dilaton. Bosonic fields without any index of spacetime is called scalar fields, so the dilaton is a kind of scalar fields.

Besides the massless particles, from the oscillation modes of a string, infinite kinds of massive particles appear. Their masses are at least $m^2 = 1/l_s^2$, and appear periodically, namely, m^2 has to be an integer multiple of $1/l_s^2$. If we look at the mass of elementary particles of our world, for example the electron and the quarks, appearing in the standard model of elementary particles, we cannot find the periodicity. Therefore, to claim that string theory describes the real world, it is natural to consider that $1/l_s$ is a considerably large mass scale so that particles with such large masses are too heavy to observe. Namely, the particles we observe are approximately regarded as massless. This huge scale $1/l_s$ can be considered to be related to the spontaneous symmetry breaking of grand unified theories, since at that scale we can see the beautiful structure of string theory. Therefore, the energy scale of the symmetry breaking may be close to the mass scale $1/l_s$ of string theory. However, even though we take $1/l_s$ to be so large, it is obvious that superstring theory still have big problems in the comparison to the real world. First, we do not live in 10-dimensional spacetime but in 4-dimensional spacetime. Second, the particles we observe are not only gravitons and photons (electromagnetism) but also various other particles. These two big problems can be solved by a method "compactification" in string theory. Next, I will explain what is the compactification which changes the dimension of the spacetime.

3.2.3 Compactified Spacetime

How should we transform the 10-dimensional spacetime which was derived due to the consistency of string theory to the 4-dimensional spacetime? You might answer that the "braneworld" which we learned in Sect. 3.1 can be brought into string theory. That is right, however, before using it we have to make clear what is the soliton realizing the braneworld in string theory. Since it, in fact, is the D-branes which are the main subject of this book, I will leave the story until Chaps. 4 and 6. Instead, here I will have an account of the "compactification" which was developed before the D-brane era as a method to transform the 10-dimensional spacetime to the 4-dimensional spacetime. Even at present times when D-branes play a crucial role, the compactification is a main research subject of string theory and is a basis on which D-branes are considered.

The compactification is a method of "making the spacetime round." As a simple example, we consider a compactification of a 2-dimensional spacetime to a 1-dimensional one. Let us imagine a paper of infinite size as a 2-dimensional space. If one makes this paper round, the cross section is a circle. If one makes it round more tightly (Fig. 3.7), the radius of the circle becomes smaller and smaller. Let us suppose that it becomes too small to see with our eyes, at the last. We can say that the dimension becomes "approximately" one dimension form the two dimensions. The reason why I emphasize the "approximately" is that, if you observe it with a good effort, you can find that in reality it is 2-dimensional. What does it mean to "observe it with a good effort" in terms of physics? The original root of this idea is old. To be more precise, this is called a Kaluza–Klein compactification, and Einstein used this to look for unified theories. The 10-dimensional spacetime with which the string theory starts is spanned by $x^0, x^1, \cdots, x^9$, and the 6-dimensional space spanned by $x^4, \cdots, x^9$ is unnecessary for us, so we will consider a compactification of this. The process of making the paper round I mentioned earlier is written in term of a mathematical equation as

$$x^i \sim x^i + 2\pi R_i \quad (i = 4, 5, 6, 7, 8, 9). \tag{3.8}$$

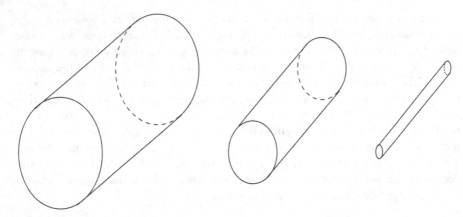

Fig. 3.7 A paper is made round and shrinks. For smaller radii, the paper which has a surface of 2-dimensional space looks 1-dimensional

Here, "$\sim$" means an identification of the two values of the coordinate. Now, the sentence "If the radius R_i is small enough, the spacetime looks 4-dimensional." does not make sense physically unless we describe what the radius is smaller than, because it has a dimension of length. I will have a more detailed explanation of this. In string theory, the gauge particles and the graviton appear as seen before. When we describe these particles by field theories, they become theories of gauge fields and a gravity field. As an example, let us consider a compactification in a field theory of a scalar particle appearing as a massless mode of closed string theory. Among the arguments x^μ of the scalar field $\phi(x^\mu)$, the 6-dimensional spacetime part x^4–x^9 is compactified by the mathematical relation (3.8). From this, the scalar field is considered to have a periodic boundary condition along the directions and can be Fourier-decomposed:

$$\phi(x^\mu) = \sum_{s_4,\cdots,s_9 \geq 0} \left[\phi_{s_4,\cdots,s_9}(x^0, x^1, x^2, x^3) \prod_{i=4,5,\cdots,9} \cos\left(\frac{s_i x_i}{R_i} + c_i \right) \right]. \tag{3.9}$$

Here, $s_4, \cdots, s_9$ are non-negative integers and c_i is a constant phase. Namely, as seen from the 4-dimensional spacetime after the compactification, the single scalar field living in the 10-dimensional spacetime turns out to be decomposed into infinite number of 4-dimensional scalar fields labeled by $\{s_4, s_5, \cdots, s_9\}$. Let us consider the masses of the these decomposed scalar fields. The equation of the motion for the massless scalar field in the original ten dimensions is

$$\partial_M \partial^M \phi = 0 \quad (M = 0, 1, \cdots, 9) \tag{3.10}$$

and if one substitutes the above decomposition equation (3.9),

$$\left[\partial_\mu \partial^\mu - \sum_{i=4}^{9} \frac{s_i^2}{R_i^2} \right] \phi_{s_4,\cdots,s_9}(x^0, x^1, x^2, x^3) = 0 \quad (\mu = 0, 1, 2, 3). \tag{3.11}$$

Therefore, these decomposed particles respectively have the following masses

$$m^2 = \sum_{i=4}^{9} s_i^2 / R_i^2.$$ (3.12)

This is the spectrum of the particles after the compactification. The lightest particle in the 4-dimensional spacetime is the unique massless scalar particle, which has the label $s_4 = \cdots = s_9 = 0$. All the others are massive scalar particles, and the lightest among them has the mass $1/R$ where R is the largest radius among the six radii compactified. Now, let us suppose that the world of our 4-dimensional spacetime is obtained in fact by a Kaluza–Kline (called "KK" in abbreviation) compactification of higher dimensions. How can we observe it? When we engage in some particle collision experiment with a particle accelerator, we can observe more new particles when the collision energy gets higher, because we gradually reach the energy to create the new particles. If the collision energy does not reach it, the particles are not created, and so they are not observed in a reality. In our present case, if the acceleration energy at a particle accelerator does not reach $1/R$, we cannot create the particles which are associated with the KK compactification (called "KK particles"). On the other hand, once this energy is reached, one can observe the KK particles by various interactions or pair creations of particles and anti-particles. Therefore, the sentence "the spacetime looks like four dimensions or not" which we used above means, in this sense, whether the KK particles characterizing the compactification can be observed or not. When we say "if you observe it with a good effort, you see ten dimensions," the "good effort" means that we increase the energy of the accelerator and observe the KK particles. It was pointed out that these KK particles may be observed at particle accelerators which will be built/operated in the future.[8] If we can accelerate particles at extremely high energy and let them collide to observe the KK particles pair-created as a result, it is a direct proof for that we actually live in a high-dimensional spacetime! I will have a full explanation of this attractive experimental observation in Chap. 6.

3.2.4 Compactification and Unified Theory

One of the problems in string theory is, aside from the problem that our spacetime is not 10-dimensional but 4-dimensional, that the actual elementary particles are not only gravitons but also other various particles. Then, how does the second problem relate with the compactification which solves the first problem? Previously we considered a scalar particle, but let us replace it with a graviton. By the KK

---

[8]In 2008, a particle accelerator called Large Hadron Collider (LHC) started to operate, and in 2009 it reached the energy which human being has never reached in history, and will reach higher energy in the near future.

compactification, massive particles labeled by non-negative integers appear, which is same as before. However, the gravity field $g_{MN}(x)$ has the indices, which is different from the scalar field. The index M and N run from 0 to 9 because the field lives originally in the 10-dimensional spacetime. (Although I wrote the index as μ so far, I will write it as M and N in order to show the sense of the higher dimensions). However, if one sees them from the 4-dimensional spacetime, the indices from 4 to 9 among them are not regarded as spacetime indices, so they are just labels. For example, the components of the gravity field of the ten dimensions $g_{\mu 9}(\mu = 0, 1, 2, 3)$ are no longer a gravity field, but a vector field (a gauge field) of the 4-dimensional spacetime! In the same way, the component g_{99} becomes a scalar field of the 4-dimensional spacetime. In this way, by the reduction of the spacetime dimensions due to the spacetime compactification, the higher-dimensional vector and tensor fields generate vector and scalar fields which have less indices of the spacetime if one sees them from the dimensions after the compactification. Using this, even though there are less kinds of particles in higher dimensions, many kinds of particles could be given in lower dimensions. If we use this idea, in string theory, we can generate many kinds of particles after the 4-dimensional compactification, in spite of smaller number of massless fields in the 10-dimensional spacetime. Then we can build "models" which are close to the actual elementary particles. Although we considered here only the simplest case of rolling a circumference, we may bring not the circumference but various 6-dimensional manifolds and regard them as the directions of the compactification. By various characteristics of the manifolds, the species of the massless particles in the 4-dimensional spacetime are determined. In late 1980s, by using this method, compactifications with various manifolds were examined, and some compactification models of string theory which reproduce the actual particle constitution were constructed. It is very interesting to explain the actual various particles unifiedly in terms of geometries of internal space, by the compactification of high-dimensional spacetimes. String theory has given a stage where this idea plays an important role, and so is paid an attention as a unified theory.

In the previous chapter and this chapter, I described what solitons are in field theories and what is string theory. We learned that solitons are important objects solving problems in field theories. And we learned also that higher-dimensional spacetimes appearing from dimensions of solitons and string theory produce various physics. In the next chapter, D-branes appear on stage finally, and it turns out that D-branes are solitons of string theory. Confirmation of this fact is a brilliant start point of physics theory of higher-dimensional spacetimes, and from there, various new theories of physics for elementary particles and cosmology are born and developed.

Chapter 4
D-Branes

As I mentioned a little in Chap. 1, D-branes are objects defined as a space on which end points of a string can be attached. Why is such a space important? For a preparation for answering this question, we have learned the solitons and their importance, and the construction of string theory and the compactification of higher-dimensional spacetime, in detail in Chaps. 2 and 3. In this chapter, based on those, I will show that the D-branes are solitons of string theory, and I will explain the importance of the D-branes. First, in Sect. 4.1, we consider what the solitons of string theory should be like, and after all we will find that they turn out to be "black holes" which are holes of spacetime. The black holes are objects whose existence is shown in Einstein's general relativity. In the sense that even light cannot escape due to too large gravity force, they are holes in spacetime. I will explain also the black holes later, but in fact the black holes are solitons, because they are defined as solutions of equations of motion of gravity. Since string theory contains gravity, solitons of string theory turn out to be the black holes. However, since string theory has higher-dimensional spacetime, the black holes appearing there show various dimensions. Here, we will see how higher-dimensional black holes appear in string theory. And, in Sect. 4.2, finally I will introduce the D-brane as a space on which the end points of a string can be attached. As a result of the definition, surprisingly, it turns out that D-branes can be identified with the black holes, and thus be solitons of string theory. And furthermore, we will see that in string theory there may exist a "duality" which is a symmetry exchanging the black holes and the strings. That is, the D-branes can be exchangeable with the string, and might play a role as a fundamental constituent of string theory. What is the ultimate theory describing all the particles and interactions in this world? As a matter of fact, string theory, which is a candidate for that, may be constructed by the D-branes. I will explain the path to the ultimate theory in Chap. 7. In Sect. 4.2, first I will start with the explanation of what D-branes are, and then of the role played by the D-branes as solitons in string theory, and finally about the duality.

K. Hashimoto, *D-Brane*, DOI 10.1007/978-3-642-23574-0_4,
© Springer-Verlag Berlin Heidelberg 2012

4.1 Soliton of String Theory is a Black Hole

In Sect. 3.2, I explained that the spacetime appearing in string theory is higher-dimensional and how our elementary particle physics of the 4-dimensional space-time can be born from that. On the other hand, in Sect. 3.2, we learned that by generalizing the solitons of the ϕ^4 model to higher dimensions, the interesting idea, the brane world, came out. In this way, you can make sure that field theories in higher dimensions and solitons there give us very interesting physics and also that string theory is a basis which provides us with the higher-dimensional spacetime in a consistent manner. In this Sect. 4.1, we shall answer the question naturally occurring from that, "what is the soliton of string theory?" It, in fact, turns out to a hole of the spacetime called a black hole.

Because string theory has a higher-dimensional spacetime, we have to generalize the notion of solutions of equations of motion of the gauge theory and the gravity theory, into the high dimension, in order to understand the solitons there. As I will describe later, gauge fields generalized to higher dimensions appear from the massless part of the string oscillations. In this chapter, I will have an explanation that the soliton of string theory is a black hole having electric and magnetic charges generalized to higher dimensions. And it will turn out that it is the very D-brane, in Sect. 4.2.

4.1.1 Strings and Charges of Tensor Fields

In order to see what gauge fields are in higher dimensions, let us review the relation between electrons and electromagnetic fields in a 4-dimensional spacetime which we are familiar with. The electron is a point-like particle, and the motion is written by the worldline $X^\mu(\tau)$. Because the electron has an electric charge, in some electromagnetic field, it moves under the influence of it. Let us write the interaction term concretely. As in (2.1) the electric field is written by the gauge field $A_\mu(x)$, and the basic idea of gauge theories is to demand a gauge symmetry that the theory has to be invariant under any gauge transformation. The gauge transformation is written as (2.14), so the interaction term we consider should be invariant under this transformation.

Such a term is written as

$$\int d\tau \, \partial_\tau X^\mu(\tau) A_\mu(X(\tau)). \tag{4.1}$$

It is obvious that this expression is invariant under (2.14). It is because, since the infinitesimal change is written by a total derivative with respect to τ, it vanishes after the integration of τ. Here, the electromagnetic field, namely, the gauge field, is written as a "background field." That is, this interaction term gives us how the electron moves in a given background electromagnetic field. In the following, I will generalize this to a higher-dimensional one, where a key point is how this interaction term can be written at the higher dimensions.

Among the massless modes of the closed string seen in Sect. 3.1, there is a Kalb–Ramond field, in addition to the gravity field. This is a field with two antisymmetric indices of the spacetime, and is written as $B_{\mu\nu}(x)$ ($B_{\mu\nu}(x) = -B_{\nu\mu}(x)$). The Kalb–Ramond field is in fact a gauge field which is like the electromagnetic field, and the gauge transformation is generalized as

$$B_{\mu\nu}(x) \to B_{\mu\nu}(x) + \partial_\mu \Lambda_\nu(x) - \partial_\nu \Lambda_\mu(x). \tag{4.2}$$

The transformation parameter $\Lambda_\nu(x)$ is now a vector field. While the electric field is sourced by the electrons, what gives the source of the Kalb–Ramond field in this string theory? We should know the answer by considering what corresponds to (4.1). It is written as

$$\int d\tau d\sigma \; \epsilon_{\alpha\beta} \partial^\alpha X^\mu \partial^\beta X^\nu B_{\mu\nu}(X) \tag{4.3}$$

(ϵ is a completely antisymmetric tensor and $\epsilon_{01} = -\epsilon_{10} = 1$.) Here, I parameterize the worldsheet by two coordinates (τ, σ) and write $\tau = \sigma^0, \sigma = \sigma^1$ as the index of this σ^α is supposed to be $\alpha, \beta = 0, 1$. This term is shown to be invariant under (4.2), too, in the same way. The important point is that in order to write the interaction term having the gauge symmetry it is necessary that it should be an interaction in $1 + 1$ dimensions. That is, the source needs to be not a particle but a string, and the string itself has an electric charge of the Kalb–Ramond field. The objects having the electric charge of the gauge field with two indices of the spacetime like the Kalb–Ramond field must have a worldvolume of $1 + 1$ dimensions.

If we apply this to higher ranks, the case of a gauge field with n anti-symmetric indices, it turns out that the object having the electric charge of it must be an extended object having the worldvolume of n dimensions (including the time dimension). (Let us see Fig. 4.1.) As we will see from now on, in string theory gauge fields with higher ranks like this in fact appear, and the higher-dimensional objects which are the source of those are the solitons of string theory, namely, D-branes.

4.1.2 Equation of Motion of String Theory

Solitons are solutions of equations of motion in field theories. Then, before asking the question "what is the soliton of string theory?", we need necessarily "string field theory."[1] What is the string field theory? In Sect. 3.1, I described that a single string

---

[1] Actually, there are two kinds of the "equation of motion of string theory," and only one of them is important for our solitons. To understand this situation, first you may think of a particle theory. For a description of particle(s), we have the one for a one-body problem and a many-body problem. The former is a worldline described by $X^\mu(\tau)$, and it is a method used for the case of a motion of one particle. The latter is, for example if it is for photons, described by a gauge field $A_\mu(x^\nu)$. Now

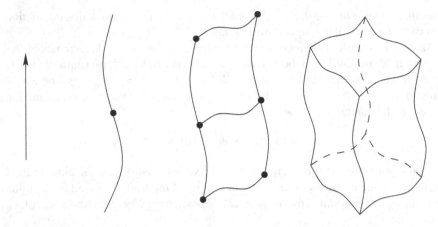

Fig. 4.1 Worldvolumes of various-dimensional objects in a spacetime. The *arrow* stands for the direction of time in the target spacetime. *Left*: a worldline of a point-like object It can be a source of a gauge field $A_\mu(x)$. *Middle*: a worldsheet of an object extending in one spatial dimension like a string. It can be a source of $B_{\mu\nu}(x)$. *Right*: a worldvolume of an object extending in two spatial dimensions like a membrane. This can be a source for a gauge field having three indices $C_{\mu\nu\rho}(x)$

is the same as infinite kinds of particles. Therefore, it turns out that the string field theory is described as a summation of actions of the infinite field theories. To obtain a soliton solution of the equations of motion of this string field theory is considerably laborious! That is because they are infinite number of differential equations mixing infinite number of fields.

In order to evade from this problem, let us make an approximation to consider the equation of motion only at the low energy region. As we saw in Sect. 3.2, unless the energy reaches $1/l_s$, the massive modes among the string oscillation modes are not excited. Namely, in this low energy approximation, one can consider only the massless particles in the spectrum. Since in string theory there are only finite number of massless particles, our problem becomes very simple.

Here as an example, we consider "type IIB superstring theory" which is one of the closed string theories having the supersymmetries. Let us write only the fields of massless bosonic particles (fields in the 10-dimensional spacetime spanned by all the x^M) among the string oscillations:

- Gravity field $g_{MN}(x)$, Kalb–Ramond field $B_{MN}(x)$, dilaton field $\phi(x)$
- Ramond–Ramond field $C(x)$, $C_{MN}(x)$, $C_{MNPQ}(x)$

---

if we look back what the soliton is in Sect. 2.3, we know that it is a solution of equations of motion of the very latter many body problem, namely, the solution of a field equation depending on x. You can easily understand this if you remember the viewpoint that solitons are a collective motion of elementary particles. Then, what is the equation of motion of string theory? There are two answers for this question, as in the same way, and the first one is equations of motion controlling $X^\mu(\sigma, \tau)$ of the one-body problem. However, this is not the answer we want for the soliton. The second, the action of the many-body problem, is important. It is the string field theory.

The newly appearing one is the Ramond–Ramond field.[2] This originates from the supersymmetry, so this doesn't appear in the case of the bosonic string. All the indexes are anti-symmetrized, for the case of the Ramond–Ramond field, and it is a gauge field generalized to higher dimensions in the same way as $B_{MN}(x)$.

As we have a complete set of massless bosonic fields appearing in the string oscillation, let us consider the action of those fields. Since the superstring theory has supersymmetries even of the 10-dimensional target spacetime, actually the action of those fields is determined uniquely. It is the action of a theory called "supergravity." In the present case, it is called type-IIB supergravity theory, from the type of the supersymmetry.[3] Here let us write only a part of it to show how the action looks.

$$S = \int d^{10}x \left[\sqrt{-\det g_{\mu\nu}}\, R[g_{\mu\nu}] + \cdots \right]. \tag{4.4}$$

This term is called the Einstein–Hilbert action, and is an action of a gravity theory. R is the quantity called Ricci scalar. Though I will not describe its detail, it is a quantity written by $g_{MN}(x)$ and its derivatives. It is an analogue of a mixture of the first and the second (potential) term of the action (2.7), in the ϕ^4 model of Sect. 2.3. String theory is, not only a theory including the rank-2 symmetric tensor field $g_{MN}(x)$ as an oscillation mode, but also a theory at which the mode obeys the same interaction as the standard gravity theory, except for the difference in dimensions of the spacetimes.

4.1.3 Black Hole

Now, let us consider, in string theory, solitons defined as solutions of the equations of motion. The equations of motion of string theory contains the gravity field, and in general, solutions of the equations of motion of the gravity field with localized energy are "black holes." Namely, it turns out that the soliton of string theory is the black hole!

---

[2]The Kalb–Ramond field was named from two scientists, Kalb and Ramond, on the other hand in the case of the Ramond–Ramond field, it does not mean there are two Ramond's. It was named from the fact that both the right- and left- movers of the fermion on the 1+1-dimensional worldsheet satisfy the boundary condition of the Ramond type.

[3]We can obtain the field action of particles corresponding to various string oscillation modes by calculating scattering amplitudes of string theory. (In the case of ordinary field theories of particles, first as a starting point we give an action of a field by hand, and then we draw Feynman graphs to compute scattering amplitudes. However, in fact in string theory, the process is opposite.) In string theory, by deforming the 2-dimensional worldsheet, we can find the scattering amplitudes of multi-strings. So as to reproduce the scattering amplitudes obtained in that manner, one can write an action of the corresponding particle fields (the correspondence can be seen as in Fig. 1.3.) In this way, in string theory, an action obeyed by the massless field like the gravity field having appeared above is written. The action obtained in this way is known to be equal to the action determined by the supersymmetry.

Here I will explain briefly what black holes are. Einstein's gravity theory, that is, the general relativity, is a theory of the gravity field[4] $g_{\mu\nu}(x)$ (it is also called "metric") and the action is given in fact by (4.4) (while the spacetime dimensions are not 10 but 4.) There is a solution of the equations of motion derived from this, which is called a black hole. The simplest one is called Schwarzschild solution, which assumes a spherical symmetry about the center of the black hole. The gravity field $g_{\mu\nu}(x)$ of the solution is written with the distance from the center $r = \sqrt{(x^1)^2 + (x^2)^2 + (x^3)^2}$ as

$$g_{00} = -\left(1 - \frac{2Gm}{r}\right), \quad g_{rr} = \left(1 - \frac{2Gm}{r}\right)^{-1}. \qquad (4.6)$$

g_{rr} is a component of the gravity field along of the radial r direction after a coordinate transformation (for the coordinate transformation, see the footnote 4). m is the mass of the black hole, and G is the coupling constant of the gravity theory (the gravitational constant). We know from this expression that, first, when r is large enough, namely, when we are very far away enough from the center of the black hole, the gravity field coincides with that of a flat (not curved) spacetime,

$$g_{\mu\nu}(x) = \eta_{\mu\nu} = \begin{pmatrix} -1 & 0 & 0 & 0 \\ 0 & 1 & 0 & 0 \\ 0 & 0 & 1 & 0 \\ 0 & 0 & 0 & 1 \end{pmatrix}_{\mu\nu}. \qquad (4.7)$$

This is the metric invariant under Lorentz transformations, which appears in special relativity. However, if we see smaller values of r, the difference from the flat spacetime metric becomes bigger. And at the radius $r = 2Gm$ the metric diverges. This radius is called the Schwarzschild radius. Inside of this radius, the spacetime is curved so strongly that one needs a velocity faster than the speed of light to escape from it. In relativity, there is no object/particle which travels faster than the speed of light. Therefore, this means that near the black-hole there is a region from which any particle cannot escape. The spherical surface defined by the Schwarzschild radius

---

[4]The physical meaning of the gravity field (metric) is as follows. Given the gravity field $g_{\mu\nu}(x)$, by using a vector (dx^μ) connecting two arbitrary points separated infinitesimally, the "proper length" ds between the two points is given as $ds^2 = g_{\mu\nu}(x)dx^\mu dx^\nu$. This proper length is a "physical" length which is invariant under any general coordinate transformation, namely, arbitrary relabeling of the coordinates, $x'^\mu = x'^\mu(x)$. This can be understood by the fact that the coordinate transformation for the gravity field is given by

$$g'_{\mu\nu} = \frac{dx^\rho dx^\sigma}{dx'^\mu dx'^\nu} g_{\rho\sigma}. \qquad (4.5)$$

The general relativity is a theory with the principle that physics is invariant under this relabeling of the coordinates, the gravity field is a basic field of that. You may understand the physical meaning of the gravity field from this proper length ds. For instance, to multiply the gravity field by a constant means that the physical distance between two points is multiplied by the constant.

is called an "event horizon" or abbreviatedly called just "horizon." Particles having entered the horizon cannot come back to the outside forever. The black hole is "a hole in the space."

Since classic solutions of equations of motion are called solitons, the solitons of the gravity theory should be the black holes. Note that in the gravity theory there is a little bit subtle point on that standard solitons are solutions of just equations of a motion. For example, the Schwarzschild solution has a problem at $r = 0$. (Although it appears that there is a problem at the Schwarzschild radius that the metric diverges, in fact we can remove this infinity by a coordinate transformation and so it is not a problem. It is just an event horizon.) The divergence at $r = 0$ cannot be removed. This point is called a singularity, where the gravity is infinite, and so any physical law at the singularity doesn't make sense. For the soliton (the kink solution) of the ϕ^4 model in Sect. 2.3, there is no such problem at its center, while for the black hole solution something strange happens there. It must be something heavy with a mass producing the gravity field. This situation is similar to the situation, for example, in electromagnetism in which when one find an electric field around an electron the electric field is divergent at the place of the electron. The electron has an electric charge and is a source of the electric field. In the gravity theory, the mass corresponds to the electric charge, and at the center of the black hole something massive which is a source of the gravity should exist as a singularity. That is, in a strict sense, black holes are not solitons. This is because they do not satisfy equations of a motion at the singularity. However, here in a broad sense, namely, in the sense that they are classical solutions except for the singularity point, let us understand the sentence that "black holes are solitons."[5]

4.1.4 Soliton of String Theory = Charged Black Brane

The low energy region of string theory is described by the supergravity theory in ten dimensions. Then, how many kinds of black holes exist as solutions of the equations of motion of this supergravity theory? As a matter of fact, the whole number is huge once we take into account the ways of the compactification, and even now they have not been classified yet. However let us remember the supersymmetries in the 10-dimensional target spacetime in this theory. Black holes which do not

---

[5]This claim that the black hole is a soliton may be correct in a strict sense if we exactly calculate it in string theory in fact. String theory becomes a supergravity theory at low energy. However, in the region near the singularity where fields vary quite rapidly, one cannot take the low energy approximation. If we treat this without the low energy approximation in a proper way in string theory, the singularity of this black hole may be actually "resolved" and the singularity might go away. There various massive fields, as well as the massless field such as the gravity, could take complicated configurations. The appearance of the singularity may be due to that we approximately focus on only the low energy of the string theory such as the gravity theory and ignore the massive fields. Therefore, the black hole could be exactly an soliton if we look at it in the whole string theory. For this, some evidence is known in string theory.

break the supersymmetries to some extent (which are called BPS black holes[6]) show some good properties, and have been studied in detail. The most important property of the BPS black hole is that it has an electric charges and the magnitude of the electric charges is equal to its mass. (Here the dimension of the mass is taken to be equal to that of the electric charge by multiplying some appropriate powers of the gravitational constant.)

The electric charge of the BPS black hole is the charge of gauge fields appearing in the supergravity theory. The gauge fields are Ramond–Ramond fields $C(x)$, $C_{MN}(x)$ and $C_{MNPQ}(x)$ in addition to the Kalb–Ramond field $B_{MN}(x)$. As I mentioned at the beginning of this section, the number of the tensor indices of the gauge field directly corresponds to the dimension of the worldvolume of the object having the electric charges. The electric charge of the Kalb–Ramond field $B_{MN}(x)$ is carried by a string with 2-dimensional worldsheet. On the other hand, the object carrying the electric charge of the Ramond–Ramond fields which newly appeared, is the one with a 2-dimensional worldvolume for example for the case of C_{MN}, and the one with a 4-dimensional worldvolume for the case of C_{MNPQ}.[7]

As the dimension of the worldvolume of the objects having the electric charges is large in this way, the electric charges of the gravity field, namely, the mass, is distributed to the extent of the worldvolume, since the energy source is accompanied with the electric charge. Namely, the singularity is not a point but a higher-dimensional object which is as large as the worldvolume. For this kind of situation, the name "black hole" is no longer appropriate. The "hole" is a point-like and so has an impression of having only one-dimensional time if one counts the spacetime dimension of the worldvolume. Instead of that, the solutions of the gravity theory with the extended mass distribution as the worldvolume are called "black p-branes." "Brane" came from a part of "membrane" and p shows the dimensions of the spatial directions of the worldvolume. So the worldvolume of the black p-brane has $p+1$ dimensions, with the time dimension. Using this labeling, the type IIB supergravity theory has BPS black 1-branes and BPS black 3-branes. There are two kind of black 1-branes due to the existence of the two kinds of rank-2 tensor gauge fields ($B_{MN}(x)$ and $C_{MN}(x)$). These are standard solitons of string theory, from the viewpoint of the low energy.

In order to write correctly the black branes as gravity solutions generated by extended singularities (with higher dimensions), we can directly apply the method to lifting the dimensions of solutions which was described in Sect. 3.1, in the same way. For instance, the Schwarzschild black hole solution in the 4-dimensional spacetime can be lifted up into 10-dimensions as it is, and it is a black 6-brane

---

[6]The "BPS" means the capital letters of three scientists, E. Bogomol'nyi, M. Prasad and C. Sommerfield. The BPS method, which was originally invented as a convenient way to find a monopole solution of a certain gauge theory, turned out to be related with the supersymmetries in later days.

[7]This applies to $C(x)$ too, and the object has a 0-dimensional worldvolume which is a point in the spacetime. This is called an "instanton." Although the instanton is a very important physical concept, I will not deal with it in this book.

solution, as understood from its dimensions. (Since this solution doesn't have the electric charge and is not BPS, it is not on the list above.) The various BPS black brane solutions listed above are made in that way.

4.1.5 Black Branes with Magnetic Charge

As we saw, the BPS black branes have electric charges. There exist BPS black branes with magnetic charges as well. However from the first place, what is the definition of the magnetic charges in the higher-dimensional string theory?

The Maxwell equations, which are equations of motion of the 4-dimensional electromagnetism, have the duality symmetry exchanging the electric and magnetic fields. Then, what about the case with the higher-dimensional and higher-rank tensor gauge fields appearing in string theory? By looking at this, one should be able to find what is the black p-brane having the magnetic charges.

The duality of the electromagnetism in the 4-dimensional spacetime is called "Hodge dual" and in the term of the gauge field strength (2.1) it is written as

$$F_{\mu\nu}(x) \rightarrow \frac{1}{2}\epsilon_{\mu\nu\rho\sigma} F^{\rho\sigma}. \tag{4.8}$$

Here, the index of $\epsilon_{\mu\nu\rho\sigma}$ takes $0, 1, 2, 3$ and it is completely antisymmetric (It is a tensor with $\epsilon_{0123} = 1$ for whose index any pair is antisymmetric.) By using the expression (2.2), one can immediately recognize that the Hodge dual exchanges the electric field and the magnetic field. Because of the 4 dimensional spacetime, we can define the completely antisymmetric tensor with four indices, therefore we can define the Hodge dual by using it. That is, one can define the field strength $F_{\mu\nu}(x)$ with two indices from the gauge field $A_\mu(x)$ with one index, and its Hodge dual is the gauge field strength with $4 - 2 = 2$ indices. And the electric charges of the Hodge dual is indeed the magnetic charge.

Let us consider this Hodge dual for the gauge fields coming from the 10-dimensional string theory. Suppose that the number of indices of the higher-dimensional gauge field is n, then following the same reasoning, we easily finds that the number indices of the Hodge dual gauge field is $8 - n$. The gauge field strength has $n + 1$ indices, and its Hodge dual is the field strength with $9 - n$ indices. Then in terms of the gauge field, the number of the indices is $8 - n$. From this, BPS black branes with magnetic charges must have the following worldvolume dimensions:

- Kalb–Ramond field $B_{MN}(x) \rightarrow$ Black 5-brane
- Ramond–Ramond fields $C(x), C_{MN}(x), C_{MNPQ}(x)$
  $\rightarrow$ Black 7-brane, black 5-brane, black 3-brane

Here, though the last black 3-brane appeared also in the previous case of the electric charges, this field $C_{MNPQ}(x)$ is originally defined as a Hodge self-dual field and so it turns out that it does not generate a new kind of black branes. Although monopole appears as an object corresponding to the electron via the exchange of the electric and magnetic fields in 4-dimensional spacetime, the reason why monopoles are point-like objects is that we consider the duality eventually at the 4-dimensional spacetime. As we learned here, in the case of gauge theories in generic spacetime dimensions and with generic number of indices, we have not only that the original objects are extended ones, but also that their Hodge dual objects with magnetic charges have worldvolume dimensions different from that of the original electrically charged objects.

To summarize the electric and magnetic charges, as the BPS black p-branes, we have

- $p = 1, 5$ comes from the Kalb Ramond field $B_{MN}(x)$,
- $p = 1, 3, 5, 7$ comes from the Ramond Ramond field $C(x)$, $C_{MN}(x)$, $C_{MNPQ}(x)$.

Objects with any odd spatial dimensions of the worldvolume appear, and in the case of $p = 1$ and $p = 5$ we have two kinds of black p-branes, as there are the Kalb–Ramond field $B_{MN}(x)$ and the Ramond–Ramond field $C_{MN}(x)$ respectively and also electric and magnetic charges. These are the solitons of the supergravity theory and therefore the solitons of superstring theory.

4.2   Emergence of D-Branes

In the previous section, we saw that the higher-dimensional black holes called BPS black branes emerge as the solitons of the supergravity theory which appear at the low energy of string theory. That is, the solitons of string theory are generalized black holes. As these black brane solutions have the center which are held up by the singularity extending to specific dimensions, what is this source of the gravity and higher-dimensional gauge fields? In this section, we will see that it is the D-branes, "a space on which end points of strings can be attached." And in this sense, D-branes are solitons of string theory. First in this section I introduce the definition of the D-branes, and next, I will have an explanation of the relation between the D-branes and the black branes, and finally I will give a meaning for the sentence "D-branes are solitons of string theory."

4.2.1   D-Branes: Space on Which Strings Can End

Open strings are defined by specifying their boundary conditions, and in Sect. 3.2, we considered only the free boundary condition (see (3.4)). Now, let us consider a fixed boundary condition instead of the free boundary condition. The fixed boundary condition is written as

$$X^{\mu}(\tau, \sigma)\bigg|_{\sigma=0,2\pi} = c^{\mu}. \tag{4.9}$$

Here, c^μ is a constant number. Generally, among the directions of the 10-dimensional spacetime in string theory, one may distinguish the directions of the free end from the directions of the fixed end. For example, the directions $\mu = 0, 1, \cdots, p$ are supposed to be subject to the free boundary condition (3.4) while the other directions $\mu = p + 1, \cdots, 9$ are to the fixed boundary condition (4.9). Then, as the end point of the string is fixed along the directions $\mu = p + 1, \cdots, 9$, the string cannot move along those directions. (It can oscillate but the center of it cannot move.) The surface on which the end points of the string are attached is, as you see in (4.9), defined by the equation

$$x^i = c^i \quad (i = p + 1, \cdots, 9). \tag{4.10}$$

This is a sub-space in the 10-dimensional spacetime. This is called a D-brane. The "D" of the D-brane comes from the fixed boundary condition, that is, the first letter of the Dirichlet boundary condition. The "Brane'" comes from the part of "membrane" as I have mentioned before.

As D-branes are defined in this manner, they are extended objects with $p + 1$-dimensional worldvolume in the 10-dimensional spacetime. Since we may consider that p is an arbitrary integer running from 0 to 9, the value p identifies the kind of the D-branes. The D-brane with the worldvolume dimension of $p + 1$ is called Dp-brane. (That is, the dimension at the spatial directions of worldvolume.) Although it is supposed here that D-branes are oriented along the spacetime coordinates for simplicity, of course the rotated object is also permitted. As long as D-branes are planar, orientations of the boundary conditions is not a problem. Furthermore, in (4.9) of the fixed boundary condition, although I write the one for which both the end points ($\sigma = 0, 2\pi$) of the string are on the same D-branes, one can choose different D-branes for each. Moreover, those D-branes may have different dimensions. As we allow such a freedom, we may consider various D-branes and strings ending on those. Let us see Fig. 1.5. There I draw three parallel D-branes (for drawing the picture I chose D2-branes) put separately and open strings which end on those.

In Sect. 3.2, I stated that the gauge field $A_\mu(x)$ appears as a massless state in the open string spectrum. Then, how about the case that D-branes exist, by changing the boundary condition? To begin with, for simplicity, let us consider a case that we have a single Dp-brane and both of the end points of the open string are attached on that (The situation of Fig. 1.5 Left.) This time, as massless boson particles,

$$A_\mu(x^0, \cdots, x^p) : \text{Gauge field} \quad (\mu = 0, 1, \cdots, p), \tag{4.11}$$

$$\Phi^i(x^0, \cdots, x^p) : \text{Scalar field} \quad (i = p + 1, \cdots, 9) \tag{4.12}$$

appear from an analysis of the string oscillation modes (I omit the detailed derivation). At first all we notice is that these particle fields are functions of $p + 1$ dimensional coordinates. That is, the field theory is limited on the D-brane. This originates from that the center-of-mass of the string cannot move in the directions transverse to the D-brane, that is, the ones of the fixed boundary conditions. By the

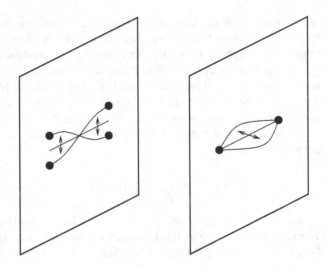

Fig. 4.2 A configuration of a string oscillation on a D-brane (a D2-brane in the figure). *Left*: the oscillation at the direction of the free boundary condition being independent of the fixed boundary condition of the D-brane. The gauge field $A_\mu(x^\mu)$ emerges from this oscillation. *Right*: The oscillation transverse to the D-brane is subject to the fixed boundary condition. From this oscillation, the scalar field $\Phi^i(x^\mu)$ emerges

same reason the gauge field is a gauge field of the $p + 1$-dimensional spacetime, that is, the index μ runs only among the $p + 1$ dimensions (Fig. 4.2).

In addition to the gauge field, scalar fields Φ appear, whose number is $9 - p$. This is the very same number for the scalar fields as the case of the dimensional reduction from the 10-dimensional gauge field to $p + 1$ dimensions (which appeared by the compactification in Sect. 3.2). As one finds that these scalar fields have the label $i = p + 1, \cdots, 9$ which are the directions transverse to the D-brane, the scalar fields are the excited modes of the string coming from the oscillation along the directions transverse to the D-brane (which is the direction of the spacetime coordinate X^i for which the fixed boundary condition is imposed). Actually, this $\Phi^i(x^\mu)$ is the field specifying the location of the D-brane in the 10-dimensional spacetime. You can understand this if you remember that the scalar field X^1 lives on the worldvolume of the kink solution and it defines the position of the kink solution (Sect. 3.1).

So far, I introduced the D-branes by considering just the fixed boundary conditions for the open string. Next, let us start with a closed string theory and introduce the D-branes. Then, we find a situation where open strings can reside only on the D-branes in the 10-dimensional spacetime, and only closed strings can propagate away from D-branes. Why do we need to introduce the D-branes in the closed string theory? The reason is that the black branes are identified with the D-branes. Next, I will have an explanation why they can be identified, but once they are identified, the D-branes have to be introduced in the closed string theory. This is because there is the supergravity theory at the low energy of the closed string theory and there are black brane solutions as its classical solutions. In this way, even though at first we start with the closed string theory, finally we have to introduce also the open strings on the D-branes.

Then, next, let us consider why the black branes are identified with the D-branes.

4.2.2 Black Branes and D-Branes

Though the D-branes are surfaces on which the end points of open strings can be attached, if we change the viewpoint of the worldsheet of the open string, we can see new characteristics of the D-branes. Let us see Fig. 1.7. In the left figure, an open string has its end on a D-brane, and it moves to form a circle as time goes. In terms of the worldsheet parameters (τ, σ), the string extends from right to left whose direction is σ, and the direction of the circumference is τ. The right figure stands for the same worldsheet, but intentionally I exchanged the roles of τ and σ. The string extends in the direction of the circumference whose direction is σ, and the string moves from the left to the right direction, and we can regard the direction as τ. Then, it turns out that this is a figure of a creation and emission of a closed string from the D-brane. The duality on the viewpoints like this is called an open-closed duality. Although first the D-branes were introduced as surfaces on which open strings can end, if we change our viewpoint of the worldsheet, we find that D-branes are sources of closed strings. Since the closed strings contain the gravity the D-branes are sources of the gravity. This is indeed the property of the black brane. Therefore, the D-branes are the black branes!

Then, how about the electric and magnetic charges which the BPS black p-branes have? Do the D-branes have the same electric and magnetic charges, too? In 1995, J. Polchinski[8] proved that D-branes are also a source of the Ramond–Ramond fields. He considered a force between two parallel D-branes as in Fig. 4.3. Because the force must be calculated by an exchange of closed strings, one may calculate the Feynman graphs like Fig. 4.3 in which a closed string moves between the two D-branes, in string theory. To calculate this graph, he used the open/closed duality which we have seen in Fig. 1.7, and regard the graph as an open string rotating once around. In this calculation of the open string, there is no need of the information of how the closed string is emitted and annihilated. Then, if we reinterpret the result as a closed string flying from left to right, we can derive the information of the magnitude of the creation and annihilation of the closed string, that is, the mass and the electric charges of the D-branes.

As a result of this calculation, he proved that the Dp-brane has an electric charge of the Ramond–Ramond field with $p + 1$ tensor indices for the case of $p \leq 3$. In addition, he showed that the Dp-brane for the case of $p \geq 5$ has a magnetic charge of the Ramond–Ramond field with $7 - p$ tensor indices. These are indeed the same as

---

[8] J. Polchinski already found D-branes in 1989 with J. Dai and G. Leigh, the importance of the D-branes has been hidden until he identified the D-branes with the black branes in 1995.

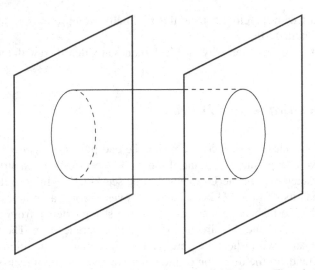

Fig. 4.3 A configuration of an interaction between two parallel D-branes. Though this looks as an exchange of a closed string in the horizontal direction, but as you can find by combining it with the previous figure, this can also be seen as an open string ending on different D-branes moves along the circumference

the black p-branes with the Ramond- Ramond charges. And, he could also showed that the calculated values of the electric or the magnetic charge of the D-brane are the same as the mass of the D-brane, which is the case for the BPS black p-branes.

As we saw in Sect. 4.1, the black p-branes with the Ramond–Ramond magnetic charges of type IIB supergravity theory are allowed to have only odd p, and the D-branes reproduce also this property. In type IIB superstring theory, if the dimension $p + 1$ of the D-brane worldvolume is not even, tachyons appear in the spectrum of the open string on the D-brane. As we saw in Sect. 2.3, the tachyons mean an instability of the field theory on the worldvolume, so the D-branes containing the tachyons are unstable. Moreover, when $p + 1$ is odd, we miss the supersymmetries in the spectrum. Namely, it does not correspond to the BPS black brane having the supersymmetries.

4.2.3 Duality Symmetry Exchanging Strings and D-Branes

What does it mean for string theory that the D-branes are solitons of string theory? Let us think about it. In Sect. 2.4, we saw that solitons called monopoles played an important role in field theories. That is, a role to solve the problem how we can calculate physical quantities in the case of large coupling constant of the field theories. When there is a "duality" exchanging a weakly coupled theory with a strongly coupled theory, the duality exchanges elementary particles with solitons. On the other hand, how about string theory? Actually, it is expected that in string theory such a duality exists. This duality is called S-duality. "S" is said to have come from the term "strong-weak," that is, strong and weak couplings.

Why do we expect that the S-duality exists in string theory? It is because the type-IIB supergravity theory has this symmetry. The whole action of the supergravity theory is invariant under the exchange of the Kalb–Ramond field $B_{MN}(x)$ with one of the Ramond–Ramond field $C_{MN}(x)$ and a simultaneous change of the sign of the dilaton field $\phi(x)$. So this is a self-duality. Let us see that this self-duality of the supergravity theory is related with the coupling constant of string theory. Once we choose the kind of the string, the coupling constant of the string theory is uniquely fixed. This is because, as you see in Fig. 1.3, whatever the oscillation of the string is, interactions among strings are unique, due to the fact that they are described by a worldsheet (for details, see Sect. 6.4). All the interactions of strings depend only on the single parameter g_s called the string coupling constant. Since there is a relation between the value of the dilaton field ϕ and the coupling constant of string theory g_s as[9] (though I will not describe the detail)

$$g_s = e^{\phi(x \to \infty)}, \tag{4.13}$$

changing the sign of the dilaton field means to make a transformation $g_s \to 1/g_s$. This is indeed a transformation to change the weak coupling to the strong coupling.

There are various other reasons we believe that the self-duality of the supergravity theory describing the low energy of string theory can be upgraded to a duality of the whole sting theory, but I will not go deeply into this subject here. The important point is that by this S-duality transformation strings and D-branes (namely solitons) are exchanged. As we saw, D-branes become sources of the Ramond–Ramond fields $C_{M \cdots N}(x)$, and D-branes are identified with the black branes with the Ramond–Ramond electric/magnetic charges. On the other hand, in Sect. 4.1, we saw that the source of the Kalb–Ramond field $B_{MN}(x)$ is given by a string itself. And by this S-duality transformation, we exchange the Ramond–Ramond field $C_{MN}(x)$ for the Kalb–Ramond field $B_{MN}(x)$, therefore it must exchange the string for the D1-brane! Namely, once we assume the existence of this S-duality, D-branes stand on a place which is as important as strings. This situation is indeed the same as the duality we have seen in Sect. 2.4, the one exchanging solitons (monopoles) for elementary particles.

Since the S-duality has been checked in various viewpoints in string theory, there is no doubt on its existence. Then, why don't we build string theory, not by using strings, but by regarding D-branes as fundamental constituent elements? In fact, such a quite attractive new theory has been already proposed, and I will introduce it in Chap. 7.

---

[9]In string theory, one expects that the value of the coupling constant g_s too is determined automatically in the framework of the theory. This is based on the relation (4.13) of this dilaton field. The reasoning is that once the "vacuum" of string theory is determined, the value of the dilaton field is fixed, and then the coupling constant of string theory is also determined.

Chapter 5
Dynamical D-Branes

The D-branes defined in the previous chapter are, if we take a glance only at their definition, just planar higher-dimensional membranes and subspaces in the whole space. However, as you will find in this chapter, the D-branes have various interesting physical characteristics. As we saw in Chap. 2, solitons behave like particles. The D-branes are solitons and therefore the D-branes are also objects moving and interacting in higher-dimensional space. In this chapter I will explain the mechanics of the D-branes, such as motion, intersection, merger, and creation/annihilation. These kinds of mechanics are described by a "field theory on the D-brane" and the fact opens up ways to interpret and analyze various field theories by the D-branes. Indeed, these field theories on the D-branes are the non-Abelian gauge theories which are the basis of the standard model of elementary particles having appeared in Sect. 2.4. Therefore, solitons concerning non-Abelian gauge theories are necessarily involved in the physics of D-branes, closely.

First, in Sect. 5.1, after I give the field theory on the D-branes, I will explain the basis of the D-brane mechanics, such as deformation, motion, and merger, from the viewpoint of the field theory on the D-branes.

Next, in Sect. 5.2, I will show that, when there are multiple D-branes, the field theory on the D-branes becomes a non-Abelian gauge theory. Using this, the monopoles of Sect. 2.4 are interpreted geometrically by D-branes in higher dimensions. The geometrical interpretation of solitons of field theories by D-branes has even a power to predict existence of new solitons.

In Sect. 5.3, I will describe the creation and the annihilation of the D-branes again from the standpoint of the field theories on the D-branes. Here, one can see once again the clear relation between solitons of field theories and D-branes. The spontaneous symmetry braking of the ϕ^4 model stands for the annihilation of D-branes, and the vortices, solitons of the field theory, stand for creation of D-branes.

The various dimensions and kinds of D-branes, and the fact that field theories live on D-branes, make a foundation of a new physics based on D-branes, and also link various physics to string theory. Furthermore, D-branes as solitons of string theory might become fundamental constituent elements of an ultimate unified theory. Then

K. Hashimoto, *D-Brane*, DOI 10.1007/978-3-642-23574-0_5,
© Springer-Verlag Berlin Heidelberg 2012

the unified theory is described by D-brane mechanics which will be introduced in this chapter. Namely, D-brane mechanics is an important basis in regard to the application for various physics and also to the ultimate theory. You will see that, in the explanation of these applications and the ultimate theory from the next Chaps. 6 and 7, D-brane mechanics appearing in this chapter provides new ideas and concepts vividly.

5.1 D-Branes Moving and Merging

The D-branes introduced in the foregoing chapter is, in 10-dimensional spacetime, a flat fixed surface determined by constants c^i appearing in the fixed boundary condition (4.10) of a string. However, in fact, the D-branes have quite dynamical physics such as motion, bending, intersection with other D-branes, merger, and what is more, creation/annihilation in a manner similar to elementary particles. It is due to the oscillation modes of the open string attached on the D-brane. A collection of those many strings brings us physics that D-branes move and vibrate. The flexibility of the D-brane mechanics supports the brilliant revolution of string theory after the D-branes appeared. In this section, let us see the basic mechanics of the D-branes and their physics.

5.1.1 Moving D-Branes

Various physics of D-branes basically depend on the gauge field (4.11) and the scalar field (4.12) living on the D-branes. First, let us see physics of the scalar field $\Phi^i(x^\mu)$. As I mentioned before, the scalar field stands for the position of the D-brane. The equation of motion of the massless scalar field $\Phi^i(x)$ is

$$\partial_\mu \partial^\mu \Phi^i(x^\nu) = 0. \tag{5.1}$$

This is the same as the equation of motion (3.1) of the scalar field which is the moving center point of the kink solution. Let us consider the following simple solution of this equation,

$$\Phi^9(x) = x^1 \tan\theta. \tag{5.2}$$

Here, θ is a constant. This means that the D-brane is rotated in the surface x^1-x^9 (Fig. 5.1). It is almost obvious that this solution is allowed. If we have considered the fixed boundary condition after rotating the original target spacetime coordinates, this kind of D-branes rotated a little should be permitted. In this solution, since the field Φ^9 which comes from a string oscillation is nonzero, we can have an interpretation that, after many strings with the oscillation gather on the D-brane and condense, as

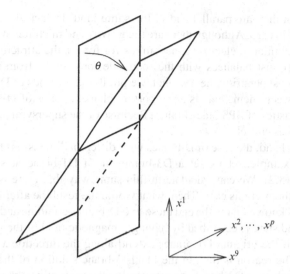

Fig. 5.1 A configuration of a D$p$brane which changes (rotates) its position in the target spacetime due to the scalar field Φ^9

a result they make the D-brane rotate. Namely, the result of the collective motion of strings gathering can be interpreted as a motion of the D-brane. This could be understood if we remember solitons of field theories, for example, the kink of the ϕ^4 model. The motion of D-branes is a gift from strings attached on them.

Then, how about the actual kinetic motion instead of the static rotation? The following solution realizes it:

$$\Phi^9(x) = vx^0. \tag{5.3}$$

Here, v is a constant. Apparently, the D-brane which this solution stands for moves at a constant velocity v to the direction x^9. On the other hand, let us consider a solution

$$\Phi^9(x) = A \cos(x^0 - x^1). \tag{5.4}$$

(A is a constant and we suppose $p \geq 1$.) This shows that a wave is generated on the surface of the D-brane and propagates along the direction x^1 in the worldvolume of the D-brane.

5.1.2 D-Brane Merger

Next, let us see how the gauge field $A_\mu(x)$ on D-branes appearing in (4.11) gives us interesting physics. This is related with a D-brane merger. In the foregoing calculation by Polchinski (Fig. 4.3), the force acting between two D-branes is calculated when they are parallel and of the same kind. In fact, the calculation of this force result in zero. Although there are the gravity and the force of the Ramond–Ramond field (similar to electromagnetism) as for forces, the attractive force made from the gravity just balances with the repulsive force made from the Ramond–Ramond field. In superstring theory, when we put the same kinds of D-branes, there is no force between them. This is an important characteristics of D-branes, and is also a characteristics of BPS black branes. Because of the supersymmetry, the forces between D-branes cancel.

On the other hand, the case of D-branes with different dimensions has a different story. As an example, let us set a D3-brane and a D1-brane at some distance parallelly (Fig. 5.2). We can calculate in this same way that there is an attractive force between them in this case. Then, what would the result be after they approach each other? It is known that in the end these two D-branes combine and form a bound state. The bound state is described by having a magnetic field of the gauge field on the D3-brane (if the original D1-branes extend along the direction x^3, $B_3 (= F_{12})$ is non-zero.) The reason is that, in the black 3-brane solution of the supergravity theory, if we consider a D3-brane as a singularity, the magnetic field on it appears to be a source of the Ramond-Ramond field $C_{MN}(x)$ as seen far from the singularity.

Furthermore, by a similar argument, if we consider an electric field $E_3 (= F_{03}) \neq 0$ along the direction x^3 in the D3-brane, in the supergravity theory it looks as a source of the Kalb–Ramond field. Therefore, the situation that there is an electric field on a D3-brane is a bound state of fundamental strings and the D3-brane. The strings of string theory are often called "fundamental strings." This is because D1-branes have the same dimensions of the worldvolume as the strings and are called "D-strings" so we need to distinguish them.) In this case the fundamental strings extend along the direction x^3 infinitely.

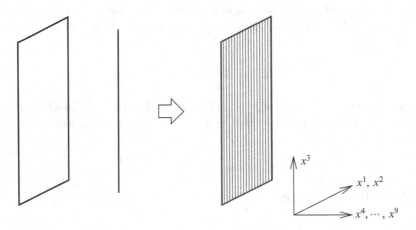

Fig. 5.2 A D3-brane and a D1-brane attract each other and form a bound state. *Left*: the D1-brane is put parallel to the D3-brane. *Right*: the two D-branes combine and form the bound state (represented by a magnetic field B_3 on the D3-brane)

The fact that the magnetic field and the electric field on the D3-brane stand for bound states is consistent with the S-duality above-mentioned. The gauge field on the D3-brane forms a gauge theory there, and it is indeed a Maxwell electromagnetism (2.3). This is because the worldvolume of the D3-brane has $3 + 1$ dimensions. Since Maxwell electromagnetism has the duality exchanging the electric field for the magnetic field (see (2.27)), after applying it to the present situation, this turns out to be the very symmetry exchanging the D1-branes of the bound state with the fundamental strings. This is the S-duality which we saw in Sect. 4.2.

Though we considered the electric/magnetic fields on the D3-brane here, this is not limited for the D3-brane. For example, let us consider one D1-brane stretching along the direction x^1 and a gauge field on that. As the worldvolume of the D1-brane has $1 + 1$ dimensions, there is only an electric field $E_1(x)$ there. If we apply the Gauss law ($\partial_1 E_1 = 0$), we find that this electric field is constant. When this electric field is non-zero, as in the case of the electric field on the D3-brane, it stands for a bound state of the D1-brane and fundamental strings attached on it. This bound state is called a (p, q)-string. p is the number of the bound fundamental strings and q is the number of the D1-branes. The (p, q)-string is a kind of D-branes, and one can find an interesting configuration of D-branes called string junctions as in Fig. 5.3.

The gauge fields on D-branes play a considerably important role, not only concerning the bound states we have seen here. We saw in the foregoing example that a single D3-brane brings us a Maxwell electromagnetism in the $3 + 1$ dimensions, but once we consider several D3-branes on top of each other, it represents a non-Abelian gauge theory. Next, let us see it. In Sect. 2.4 we learned

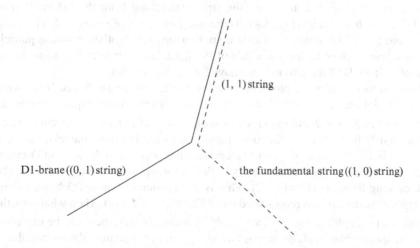

$(1, 1)$ string

D1-brane $((0, 1)$ string$)$ the fundamental string $((1, 0)$ string$)$

Fig. 5.3 A string junction. Three (p, q)-strings extending half-infinitely join together at a point and they are stable with a force balance. The *dashed line* stands for a fundamental string and the *solid line* stand for a D1-brane. The part at which two lines run parallelly is a bound state called $(1,1)$-string. At the center of the junction, strings prong in such a way to preserve the number of the fundamental strings and that of the D1-branes

monopoles as solitons of a non-Abelian gauge theory, and if we use the realization of the non-Abelian gauge theory by D-branes, the monopoles can be surprisingly described by D-branes!

5.2 Non-Abelian Gauge Theories on D-Branes and Solitons

5.2.1 D-Branes Generating Non-Abelian Gauge Theories

Let us see an interesting physics produced by D-branes of the same kind on top of each other. In the situation of two parallel D3-branes, as you see in the right Fig. 1.5, open strings connecting the two D3-branes exist. Massless particles do not appear from oscillation modes of the strings, because energy is present with stretching of the string and a positive constant is added to the mass formula. For example, in the case of a bosonic string, the mass formula of the string connecting the two parallel D-branes at some distance is modified as:

$$m^2 = \frac{1}{l_s^2}\left(-1 + \sum_{n>0} n N_n\right) + \left(\frac{\text{the distance between D}-\text{branes}}{2\pi l_s^2}\right)^2. \quad (5.5)$$

However, on the other hand, when the distance between the D3-branes becomes zero, that is, if the D3-branes are on top of each other, massless particles appear in the spectrum of the string connecting these two D3-branes. Let us name the two D3-branes for $a = 1, 2$ and write the string stretching from the D3-brane a to the D3-brane b as $\langle a, b \rangle$ (Fig. 5.4). There are four types of strings, $\langle 1, 1 \rangle$, $\langle 1, 2 \rangle$, $\langle 2, 1 \rangle$ and $\langle 2, 2 \rangle$. When the D3-branes are put on top of each other, massless particle appears from each of them. The kinds of the particles are exactly the same as the case of a single D3-brane, so now we have four copies of them.

First let us consider the physical meaning of the four scalar fields. If we write the scalar field as $\Phi^i_{\langle a,b \rangle}(x^\mu)$ $(a, b = 1, 2)$, as it is clear from explanation of the scalar field in Sect. 4.2 and the previous section, $\Phi^i_{\langle 1,1 \rangle}(x^\mu)$ stands for the position of the first D-brane (in the direction transverse to the D-brane, namely, the right direction in Fig. 5.4), and $\Phi^i_{\langle 2,2 \rangle}(x^\mu)$ stands for the position of the second D-brane. Then, the motion of the center of mass of the D3-branes is the average of the scalar field coming from $\langle 1, 1 \rangle$ and $\langle 2, 2 \rangle$. That is, the motion of whole D3-branes along the transverse direction is represented by $(1/2)(\Phi^i_{\langle 1,1 \rangle} + \Phi^i_{\langle 2,2 \rangle})$. Then what does the remaining $(1/2)(\Phi^i_{\langle 1,1 \rangle} - \Phi^i_{\langle 2,2 \rangle})$, $\Phi^i_{\langle 1,2 \rangle}$, $\Phi^i_{\langle 2,1 \rangle}$ mean? In fact, these can be identified with the three scalar fields ϕ_1, ϕ_2, ϕ_3 having appeared concerning the monopoles in the last part of Sect. 2.3. Although the scalar fields ϕ_1, ϕ_2, ϕ_3 originally contains the tachyonic negative m^2 and the interaction λ in the field theory, if we take a limit in which both m and λ go to zero (which is called a BPS limit), it is known that it becomes a theory of those three scalar fields, $(1/2)(\Phi^i_{\langle 1,1 \rangle} - \Phi^i_{\langle 2,2 \rangle})$, $\Phi^i_{\langle 1,2 \rangle}$, $\Phi^i_{\langle 2,1 \rangle}$.

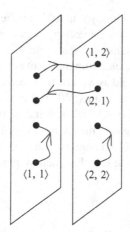

Fig. 5.4 A classification of open strings when there are two D3-branes put parallelly. The *arrow* represents the orientation of the parameterization σ. We suppose that D-branes are on top of each other. From this, a non-Abelian gauge theory appear on the D-branes

Moreover, on the D3-branes there appear gauge fields $(1/2)(A_\mu^{\langle1,1\rangle} - A_\mu^{\langle2,2\rangle})$, $A_\mu^{\langle1,2\rangle}$, $A_\mu^{\langle2,1\rangle}$, correspondingly, and they can be identified in the same light as the three gauge fields related with the rotation in Fig. 2.15 in Sect. 2.3. These fields form a non-Abelian gauge theory introduced in Sect. 2.3.[1]

It is surprising enough that non-Abelian gauge theories of $3 + 1$ dimensions appear in a very simple fashion from string theory. This is because important theories in elementary particle physics, such as the standard model of elementary particles and the grand unified theory, are non-Abelian gauge theories in $3 + 1$ dimensional spacetime. The spacetime dimensions are given by the worldvolume dimensions of the D-branes, and the gauge symmetry can be supplied by piling up the D-branes. This mechanism of D-branes gives not only a big help for trying to give the standard model of elementary particles from string theory, but also methods to study how the standard model is naturally understood from string theory and what is the grand unified theory favored in string theory. Let us explain the part of these developments in Chap. 6.

Here, we saw that, when two D3-branes are put on top of each other, there emerges a non-Abelian gauge theory whose gauge symmetry is a symmetry rotating a spherical surface in a 3-dimensional space having appeared at the last of Sect. 2.3. Let us generalize this and consider a situation where the number of the D3-branes in the pile is N. The gauge fields and the scalar fields are labeled by $\langle a, b \rangle$ $(a, b = 1, 2, \cdots, N)$ in the same way, and each number is N^2. The $N^2 - 1$ gauge fields except for the center of mass part correspond to a more generalized

---

[1]More precisely, in order to identify it with the non-Abelian gauge theory, one has to show that interactions of these oscillation modes equal to the interactions of the non-Abelian gauge theory. This can be checked by calculations of scattering amplitudes in string theory.

non-Abelian gauge symmetry. This symmetry is mathematically described by a Lie group called SU(N)[2] and this is the one generalized from the rotational symmetry of the spherical surface in a 3-dimensional space (the Lie group called SU(2)[3]). On the other hand, the symmetry of the electromagnetism is same as a degree of freedom of a phase rotation of a complex field as we saw for the vortex solitons in Sect. 2.3. This is called U(1) in terms of the Lie groups. When there is a single D-brane, we have a gauge theory of a gauge symmetry U(1). The standard model of elementary particles is described by a non-Abelian gauge theory based on the symmetries SU(2), SU(3), and U(1). (The SU(3) part is QCD having appeared in Sect. 2.4.) From that, one can understand the importance of the fact that such symmetries emerge from D-branes in a simple manner.

5.2.2 Seeing Monopoles: D-Brane Prediction of New Solitons

Among field theories, which are not only the standard model of elementary particles describing the real world but also various field theories studied in order to understand the physics of it, interesting field theories are non-Abelian gauge theories. Once D-branes construct theories in various dimensional spacetime with various symmetries, the road to study the theories from string theory opens in front of us. One example is a "visualization" of solitons such as monopoles by the D-branes. It is possible to represent monopoles by a geometrical configuration of D-branes in higher dimensions. To understand monopoles geometrically means to understand them quite intuitively. As a result, the existence of new solitons other than monopoles were predicted by the D-branes! Furthermore, amazingly, the prediction was proved to be correct.

First, in order to see how the monopoles are described by D-branes, let us consider how the spontaneous symmetry breaking occurs in the system of D-branes. As we have seen the way we construct non-Abelian gauge theories by using D-branes, there we just pile up two D3-branes on top of each other. Let us make these two D-branes separate from each other while keeping them parallel. Then, since the string $\langle 1, 2 \rangle$ and the string $\langle 2, 1 \rangle$ connect separate D3-branes, massless particles do not appear in their oscillation modes (spectra). Namely, $A_\mu^{\langle 1,2 \rangle}(x)$ and $A_\mu^{\langle 2,1 \rangle}(x)$ obtain their masses and as a result lose the gauge symmetry. On the other hand, $(1/2)(A_\mu^{\langle 1,1 \rangle} - A_\mu^{\langle 2,2 \rangle})$ remains massless and still have the gauge symmetries related with it. This situation is indeed the same as the spontaneous symmetry breaking by a vacuum condensation which was explained at the last of

---

[2]A kind of the Lie groups, SU(N), is a group made by $N \times N$ matrix M where M satisfies $MM^\dagger = M^\dagger M = 1_{N \times N}$ (the unitarity) and det $M = 1$. "SU" means the initial letters of "special unitary" ("special" stands for the condition det $M = 1$).

[3]Precisely speaking, this rotational symmetry is called SO(3). This is locally equivalent to SU(2) but globally different. In this book we ignore this difference.

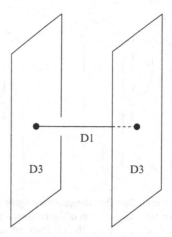

Fig. 5.5 The figure of D-branes representing a monopole. A D1-brane is suspended between two parallel D3-branes

Sect. 2.3. In Fig. 2.16, I mentioned that after the breaking the rotational symmetry around the chosen vacuum remains, and the gauge field of the symmetry is in fact $(1/2)(A_\mu^{\langle 1,1 \rangle} - A_\mu^{\langle 2,2 \rangle})$.[4] In this way, the spontaneous breaking of the gauge symmetry of the non-Abelian gauge theory is determined by the relative position of the D3-branes. This is very interesting in the sense that the symmetry breaking is geometrically visualized.

As the symmetry breakings are deeply related with soliton, we shall see that monopoles can be expressed by D-branes. To say the conclusion first, the monopole corresponds to the D-brane configuration of Fig. 5.5. This is a D1-brane that is suspended between two parallel D3-branes, and so the D1-brane has its end points on the D3-branes.

If we consider the reason why the D1-brane can have its end points on the D3-branes in this way, we may understand why Fig. 5.5 is the monopole of Sect. 2.3. First, let me remind you of the aforementioned S-duality, it is a symmetry exchanging the fundamental string for the D1-branes. And the fundamental strings can end on D3-branes. (This is the definition of D-branes.) Therefore, the D1-brane can end on D3-branes, due to the S-duality. Let us consider the meaning of the end points. We consider the following procedures, in order to understand Fig. 5.5. We put a single D1-brane on the two D3-branes on top of each other, as in the left figure of Fig. 5.6. This must be described by a magnetic field on the D3-branes. Then let us separate these two D3-brane a little, and simultaneously we keep each end point of

---

[4]On the other hand, since $(1/2)(A_\mu^{\langle 1,1 \rangle} + A_\mu^{\langle 2,2 \rangle})$ concerning the center-of-mass motion is not affected even if we change the relative position of the D3-brane, it still remains massless. So there are two remaining massless gauge fields, $A_\mu^{\langle 1,1 \rangle}(x)$ and $A_\mu^{\langle 2,2 \rangle}(x)$. These are the gauge fields localized independently on each D3-brane. Once we regard the two D3-branes as just in independent motion, we can understand easily that these two gauge fields remain massless.

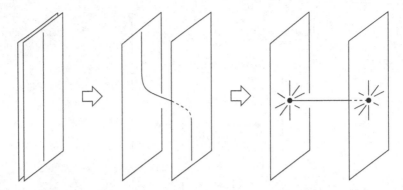

Fig. 5.6 A deformation for understanding that the previous figure represents a monopole. *Left*: we put a single D1-brane on two D3-branes on top of each other. The D1-brane is expressed as a magnetic flux in the D3-branes. *Center*: we make the D3-brane separate from the other. When the distance is large enough, the D1-brane looks suspended between the D3-branes vertically. The end points of the D1-brane turn out to be sources of the magnetic field on the D3-branes

the D1-brane remaining on a different D3-brane. The result of making them separate more is Fig. 5.5. As you may understand from this, the end point of the D1-brane must be a source for the magnetic field on the D3-branes. This is indeed the magnetic monopole. In order for the monopole to be present, there has to be a spontaneous symmetry breaking, which now corresponds to the distant D3-branes.

The visualization of the monopole by the D-brane is indeed a quantitative correspondence. For example, the mass of the monopole can also be calculated by D-branes. It is known that in the analysis of the field theory the mass of the monopole is determined by a scale of the symmetry breaking, that is, the magnitude $|\phi|$ of the field at the vacuum. (In the BPS limit, the mass is completely proportional to the magnitude ϕ in the vacuum.) The magnitude of the field at the vacuum is a square root of the right side of (2.26), while in terms of D-branes it is the distance between the D3-branes. For simplicity, I have chosen $(1/2)(\Phi^i_{(1,1)} - \Phi^i_{(2,2)})$ as a corresponding field on the D-3brane, among ϕ_1, ϕ_2, ϕ_3. Moreover since the D1-brane extends between the D3-branes, the energy of the D1-brane is proportional to the distance between the D3-branes. (The energy of the D1-brane is the energy per a unit length (tension) multiplied by the total length.) Namely, to conclude, the mass of the monopole is proportional to the energy of D1-branes. In fact it can be found that the proportionality coefficient completely coincide, by a detailed analysis. The quantitative properties of monopoles can be calculated by a geometrical configuration of D-branes.

As in this way we grasp the quantitative properties of monopoles, D-branes turn out to provide us with a technology at the level that it can be used to look for new monopole solutions. In 1997, Bergman made a conjecture that a new soliton solution should exist in non-Abelian gauge theories which looks like Fig. 5.7, by exchanging the D1-brane of Fig. 5.5 for a string junction of Fig. 5.3. Due to the structure of the string junction, three D3-branes have to be there parallelly. And the mass of the

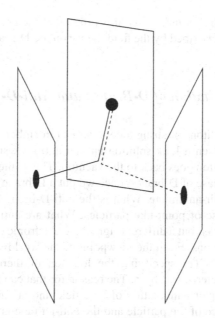

Fig. 5.7 A new kind of monopoles which was predicted by string theory. Once we replace the D1-brane in the monopole by a string junction, we have this D-brane configuration (The actual orientations are a little different in this figure as we make a reduction from higher dimensions to the 3-dimensional space.)

soliton solutions must be the energy of the string junction (which is a sum of the tension of each string multiplied by each length). This was a conjecture made by string theory for field theories. Later, corresponding equations of the motion of the non-Abelian gauge theory were concretely solved, and this interesting soliton solution was constructed exactly on the field theory side, and the existence was proved. And the mass of the soliton turned out to be identical to the energy of the string junction. Namely, the conjecture of string theory was correct.

In this manner, by applying the fact that non-Abelian gauge theories are on D-branes and soliton solutions there are also described by D-branes, we have had developments in researches of field theories. The understanding about which kinds of D-branes interact in what way in string theory reveals the properties of solitons in field theories.

5.3 Creation/Annihilation of D-Branes and Tachyon Condensation

We have seen in Sect. 4.2 that, thanks to the S-duality, D-branes might be fundamental objects of string theory instead of fundamental strings. If we suppose this is the case, the process of creation and pair-annihilation of D-branes has to

be able to be described. Let us see that this process of the creation/annihilation of
D-branes can also be described by the field theory on the D-branes in fact.

5.3.1 Pair-Annihilation of D-Branes and Anti-D-Branes

While D-branes are solitons of string theory, let us remember how solitons of field
theories annihilate. When a kink solution and an anti-kink solution collide as in
Fig. 2.8, the configuration goes back to the vacuum. This is the pair-annihilation of
solitons. Then, in the case of D-branes, once we put a D-brane and an anti-D-brane
together, they must pair-annihilate. What is the anti-D-brane? For simplicity, first
let us consider the case of point-like particles. What are anti-particles? They are
things with the same mass but a different sign of the electronic charge, like positrons
corresponding to electrons. From the viewpoint of the worldvolume, anti-particles
have the worldlines $X^\mu(\tau)$ specifying the location of them with the opposite
direction of the parameterization by τ. The reason for that can be easily understood
by the process of the pair annihilation of a particle and an anti-particle (Fig. 5.8,
left). The disappearance of the particle and the anti-particle is shown by the figure
in which a single worldline goes from the past to the future and returns back to the
past again. As for D-branes, the same reasoning leads us to the anti D-branes. The
orientation of the worldvolume in the higher dimensions is opposite. For example,
let us describe a worldvolume of a D1-brane by two coordinates, τ and σ. Supposing
that the D1-brane extends along the direction X^0 and X^1 in the target spacetime,
then we have two simplest ways of the parameterization, as

$$(1) \quad \tau = X^0, \quad \sigma = X^1, \qquad\qquad (2) \quad \tau = -X^0, \quad \sigma = X^1.$$

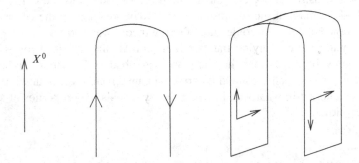

Fig. 5.8 *Left*: worldlines describing a pair-annihilation of a particle and an anti-particle. The
vertical direction stands for the time direction. At a certain instance the worldlines of the particle
and the anti-particle join, which stands for the annihilation. For this to happen consistently, the
directions of the parameterization of the worldlines must be opposite to each other. The *arrow*
placed at the left stands for the direction of time of the target spacetime. *Right*: the worldvolumes
standing for a pair annihilation of a D1-brane and an anti-D1-brane. In the same way the labelings
of the orientations are opposite to each other

This (1) is a D1-brane and (2) is an anti D1-brane. You can understand this by drawing a figure of the worldvolume giving the pair annihilation of (1) and (2) (Fig. 5.8, right). This way of thinking can be easily extended even for higher-dimensional worldvolume of D-branes.

Then what will happen when a D-brane and an anti D-brane are on top of each other? Let us look at Fig. 5.4 again. Previously, we considered two D-branes of the same kind, while now we suppose that there is a D-brane and an anti D-brane. Then what would be the difference? Among four kinds of strings, the string $\langle 1, 1 \rangle$ and the string $\langle 2, 2 \rangle$ are in the same situation as before, so they generate massless particles (4.11) and (4.12). However, $\langle 1, 2 \rangle$ and $\langle 2, 1 \rangle$ connect D-branes of opposite orientations, so the particle spectra coming from the oscillation modes are known to be different from that of $\langle 1, 1 \rangle$ and $\langle 2, 2 \rangle$. Though I omit the explanation, the mode with the lowest m^2 among the spectra is not the massless field but a tachyon field.

$$T^{\langle 1, 2 \rangle}(x^0, \cdots, x^p), \quad T^{\langle 2, 1 \rangle}(x^0, \cdots, x^p) : tachyon, \quad m^2 = -\frac{1}{2l_s^2}.$$

Precisely as in the ϕ^4 model seen in Sect. 2.3, a tachyon field appears. The point $T = 0$ is at the top of the potential and so unstable. This means that the D-branes we consider are unstable.[5]

The real vacuum places at the bottom of the potential, and there the theory should be stable. Using this way of thinking, Sen in 1998 made a conjecture as below:

• When the tachyon particle coming from the string connecting the D-brane and the anti D-brane reaches the true vacuum after the vacuum condensation, it is equivalent to the pair annihilation of the D-branes.

This conjecture was proven by an examination of the bottom of the potential of the tachyon field derived in string theory. Concretely speaking, since the potential of the tachyon field interacts complicatedly with infinite number of other oscillation modes, one must handle them all. This is the string field theory which was mentioned in Sect. 4.1. It was shown that, if one solves the simultaneous nonlinear equations of infinite dimensions containing infinite number of fields, and calculates the hight (Fig. 5.9) from the bottom of the potential to the top of it, then it is equal to the mass energy (that is the tension) per a unit area of the D-brane and the anti D-brane. Furthermore, interestingly, it is shown that all the oscillation modes of the open strings which were originally there disappear at the bottom of this potential.[6] This is exactly corresponding to the annihilation of the D-branes.

---

[5]In the type IIB superstring theory, I described that D$p$brane with odd p only exists, while when p is even, a tachyon appears in its spectrum and that stands for a D-brane which is unstable by itself. This is called a non-BPS D-brane.

[6]It is a very interesting situation that the annihilation of the D-branes themselves can be described by the strings on the D-branes. In fact, when one gets closer to the bottom, the speed of light gets smaller and in the end it vanishes. This means that all the oscillation modes cannot move. The limit

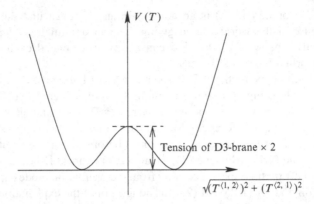

Fig. 5.9 The height of the tachyon potential. The energy difference between the vacuum and the potential top becomes the tension of the D3-brane and the anti D3-brane disappearing

5.3.2 Creation of D-Branes

As there are two tachyon fields, $\langle 1, 2 \rangle$ and $\langle 2, 1 \rangle$, you might recognize that it resembles the situation of the vortex solitons as we saw in Sect. 2.3.[7] $T^{\langle 1,2 \rangle}$ and $T^{\langle 2,1 \rangle}$ are ϕ_1 and ϕ_2, respectively.[8] Using this similarity, Sen made the second conjecture:

- The vortex soliton made by the two tachyon fields on the Dp-brane and the anti-Dp-brane is a $D(p-2)$-brane.

Here the reason why the dimensions of the worldvolume of the D-branes decrease by two is that, as we saw in Sect. 3.1, the vortex is basically a soliton localized in a two-dimensional space and, if we generalize it (via the dimensional reduction) to a p-dimensional space, the spatial part of the worldvolume of the vortex soliton turns out to has $p - 2$ dimensions. Namely, the reason is that the co-dimension of the vortex soliton is two.

For this conjecture, a proof was given by showing that the mass of the vortex soliton as a solution of equations of motion of a string field theory equals the tension of a $D(p-2)$-brane (the mass energy per a unit area).

---

in which the velocity of light equals zero is called a "Carroll limit." A mathematician L. Carroll is the author of "Alice in Wonderland" where there is a scene in which Alice feels she cannot move.

[7] In the case of the non-BPS D-brane there is only a single tachyon, and it is just like the ϕ^4 model.

[8] In the previous section, we considered two D-branes sharing the same direction, then the massless scalar fields $\Phi^i_{\langle 1,2 \rangle}$ and $\Phi^i_{\langle 2,1 \rangle}$ appear, and together with $(1/2)(\Phi^i_{\langle 1,1 \rangle} - \Phi^i_{2,2})$, they correspond to ϕ_1, ϕ_2, ϕ_3 appearing in monopoles in Sect. 2.3. The point on which we should not be confused with the story in this section is that in the foregoing chapter we took the "BPS limit" in which these ϕ_1, ϕ_2, ϕ_3 become massless. This time, we don't take this limit, and in fact one can directly identify ϕ_1 and ϕ_2 for the vortex in Sect. 2.3 with the tachyonic field $T^{\langle 1,2 \rangle}$ and $T^{\langle 2,1 \rangle}$ with the negative m^2.

In this way, the D-branes with lower dimensions generated by the pair-annihilation of D-branes are described by a vacuum condensation of the tachyon field on the D-branes. It is interesting that we can understand the creation / annihilation of D-branes as a behavior of the tachyon field living on the D-branes themselves.

5.3.3 Reconnection of Intersecting D-Branes

As an application of this annihilation of D-branes, let us consider intersecting D-branes and their reconnection. We put two D1-branes as an example, let them intersect each other, and call the intersection angle θ ($0 \leq \theta \leq \pi$). Since D-branes have orientations, the case of $\theta = 0$ means parallel D1-branes, while the case of $\theta = \pi$ shows a D1-brane and an anti D1-brane on top of each other.

Let us consider the strings $\langle 1, 2 \rangle$ and $\langle 2, 1 \rangle$ which were mentioned previously. These generate the massless gauge fields for $\theta = 0$ while the tachyon fields for $\theta = \pi$. Since these two should be connected by a continuous deformation, for generic θ, m^2 has to be between those two values. In fact, the spectrum of the string is calculated as

$$m^2 = -\frac{\theta}{2\pi l_s^2} \tag{5.6}$$

and it turns out to be the case. That is, when θ is nonzero even with a slightest angle, the tachyon field appears. Then, let us consider what happens, for an intermediate value of θ (which is neither 0 nor π), with a vacuum condensation of this tachyon field. The characteristic feature in this case of the intermediate value of θ is that the strings $\langle 1, 2 \rangle$ or $\langle 2, 1 \rangle$, that is, the string connecting the two D1-branes, exist only around the intersection point (Fig. 5.10). Since strings have tensions, they try to connect the two D-branes at the shortest distance. Then, applying the

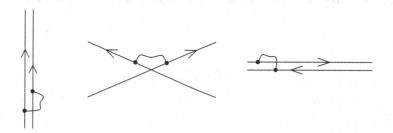

Fig. 5.10 Intersecting D1-branes (*center*) become D1-branes (*Left*) put on top of each other when the intersection angle θ is 0, while become a D1-brane and an anti D1-brane (*Right*) on top of each other when the intersection angle θ is π. When the intersection angle θ is neither 0 nor π, a fundamental string (*a wavy line*) connecting the two branes is localized at the intersection point

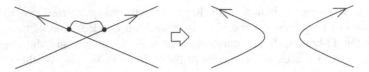

Fig. 5.11 If the tachyon modes localized at the intersection point go through the vacuum condensation, the intersecting D-branes reconnect. The reconnection at the intersection point can be regarded as a localized pair-annihilation of the D-brane and the anti D-branes

Sen's conjecture shows that the D-branes have to pair-annihilate only around this intersection point.

This localized pair-annihilation, in fact, corresponds to the reconnection of the D1-branes. This would be obvious when you see Fig. 5.11. You try to erase the D-branes only around the intersection point while keep the remaining parts un-erased, then the only possible configuration left is the reconnection. One can intuitively understand the reconnection from the viewpoint of the energy. Since the intersecting D-branes can reduce their length by the reconnection, they can reduce their energy. Therefore, the intersecting D-branes are unstable states concerning the energy. The decrease of the energy by the reconnection just corresponds to the potential energy gained by the vacuum condensation of the tachyon. In fact, the reconnection of the intersecting D-branes can be concretely shown by the non-Abelian gauge theory on the D-branes.

In the previous chapter, we saw that D-branes can be recognized as solitons of string theory, namely, black branes. On the other hand, in this chapter we saw that the physics such as motion, merger and annihilation of D-branes, can be described by the non-Abelian gauge theory on the D-branes and its solitons. These two stories apparently look irrelevant. One story is with a close connection with gravity in which D-branes play an important role as solitons in string theory, while the other story is that motion of D-branes can be described by the field theory. However, in the very fact that the D-brane exhibiting these two properties is in fact the same object in string theory, secrets of D-branes on their surprising applicability are hidden. Let us find out these secrets, in the following chapters, as I introduce various interesting applications of D-branes. One can realize the actual power of D-branes there.

Chapter 6
Application of D-Brane Physics

Since the discovery of D-branes as solitons of string theory by Polchinski in 1995, the application research has been explosively developed. This word "explosively" is not an exaggeration. D-branes have given tremendous influences not only to the framework of string theory but also to various physics around, by their flexibility. They have brought us not a slight influence but a huge revolution such as a creation of new subjects and supply of new paradigms. In this chapter, I will pick up four topics which have been considerably developed intrinsically owing to the emergence of the D-branes, and will have a brief explanation for each. As you can read these four sections independently, you might start with whichever section you are interested in. Since they are leading-edge research results, I am forced to omit details a little bit. However, I hope you may read the great influences given by D-branes, the possibility of their future, and the excitement of researchers engaged in the study of D-branes.

First in Sect. 6.1, I will explain the application of the "braneworld" which was mentioned in Sect. 3.1. The braneworld means a hypothesis that we live on D-branes. It not only is naturally expected from string theory but also gives various merits in elementary particle physics. It not only gives us a higher-dimensional interpretation of the characteristics of known elementary particles and their interactions, but also becomes a method to resolve problems of the standard model of elementary particles. Furthermore, supposing the braneworld, we can expect quite interesting experimental results. Let us see them later.

Next in Sect. 6.2, as an application of the braneworld, I will explain an understanding by D-branes of inflationary cosmologies of the expansion of our universe. The important process of the universe expansion called an inflation is interpreted by a motion of a D-brane in higher-dimensional space. The dynamics of D-branes we saw in the previous chapter is actually made the best use of in cosmology.

In Sect. 6.3, I will explain an example in which the notion of D-branes gives us a landmark outcome in the physics of black holes. Although the black holes are "space holes" which are difficult to handle in general relativity, their thermodynamic

K. Hashimoto, *D-Brane*, DOI 10.1007/978-3-642-23574-0_6,
© Springer-Verlag Berlin Heidelberg 2012

properties are revealed by D-branes. D-branes give the answer to the question : what is the entropy of black holes?

And, in Sect. 6.4, I will explain a duality called "holography" which is an application of the identification of black holes and D-branes. This is a very intriguing equivalence at which a gravity theory is equivalent to a non-Abelian gauge theory, and moreover, the dimensions of the spacetimes of these theories of our concern are different. The holography is a new correspondence brought by an understanding of the D-branes, and this is a considerably innovative method having the possibility to solve the problems of the standard model of elementary particles, such as the quark confinement, calculations of spectra of QCD, and so on.

6.1   Application I: Braneworld

6.1.1   Beyond the Standard Model of Elementary Particles

Elementary particle physics has been developed to explain theoretically new particles and their interactions, observed at particle accelerator experiments and radiations. When a theory is written to explain a certain particle and its interaction, from a consistency of the theory, some existence of a new particle or a new interactions is predicted, and then observation of them proves the validity of the theory. Repetition of this process has made the present Standard Model of elementary particles. The Standard Model is almost perfect. The only particle which has not been observed among the Standard Model particles is the Higgs particle. Except for some latest experimental results, all experimental and observed results can be explained without contradiction by the Standard Model of elementary particles.

However, in the Standard Model, there are some problems. It is considered that they are rather theoretical problems, and will be explained by a bigger theory containing (or reproduce in a certain limit) the Standard Model. One of the problems is the "hierarchy problem" which will be described below. The braneworld I will explain in this section is a new and very interesting idea to solve the hierarchy problem. The braneworld not only can solve the problem but also will predict substantial and various important physics, and they are expected to be able to be observed at future accelerator experiments. In this section, after describing first what the hierarchy problem is and what the braneworld is, I will explain how it solves the hierarchy problem, and then describe very interesting experimental results predicted by it.

6.1.2   Standard Model and Hierarchy Problem

The Standard Model does not contain the gravity. Why? The Standard Model is written as a field theory, and it is treated as a quantum field theory. As I mentioned in Sect. 2.4, in quantum field theory, corrections to the interactions and masses are calculated by the effects of various Feynman graphs (higher order effects in perturbation theory). These results containing the "quantum corrections" accurately reproduce experimental results at accelerators. If one tries to include the gravity here, it creates a problem of having the non-renormalizable infinity. When we calculate quantum corrections in the field theory of the Standard Model, a lot of infinities appear in the result of the calculations. However, we can make these infinities finite by a process called "renormalization," so they are "well-behaved" infinities. On the other hand the infinities appearing in the quantum corrections by gravity, that is, in the contribution of Feynman graphs in which gravitons fly, cannot be made finite by the "renormalization." They are "bad-behaved" infinities. String theory appeared as a theory making quantum correction of this gravity finite. Namely, string theory is a theory unifying Einstein's general relativity and quantum mechanics consistently.

In spite of the fact that the Standard Model does not contain gravity, why the Standard Model succeeds in describing the properties of elementary particles? It is because the magnitude of the gravity interaction is immeasurably smaller than that of the interactions contained in the Standard Model of elementary particles. And so there is no problem in ignoring gravity temporarily when describing results of accelerator experiments.

However, conversely, this offers one problem: Why gravity is much weaker than the other interactions? This problem is called "a hierarchy problem" in the sense that the energy scale of gravity interaction is quite separated from that of the other interaction. Why does nature provides us with hierarchical strengths of forces?

To show the seriousness of this problem, I will explain how weak the gravity interaction is. In the Newton's law of gravity, the force between the objects with mass m_1, m_2 is given as

$$F = G \frac{m_1 m_2}{r^2}. \tag{6.1}$$

Here G is called the gravitational constant. In the natural Planck unit, we take $c = 1$ and $h/2\pi = 1$ (c is the speed of light, and h is Planck constant). In this unit system, we can rewrite the gravity constant in the unit of energy (or mass) and it is called the Planck scale or the Planck mass. The value of it is about

$$M_{\text{Pl}} = \frac{1}{\sqrt{G}} = 10^{19} \quad [\text{GeV}]. \tag{6.2}$$

This unit "GeV" is an abbreviation of giga electron volt. One electron volt amounts to the energy obtained by a single electron accelerated between two plates with 1-volt difference in their electric potentials. In terms of mass it is 1 [eV] $\simeq 1.73 \times 10^{-36}$ [kg], but in elementary particle physics this eV is usually used as a unit of energy and mass. On the other hand, the energy scale of the spontaneous symmetry breaking by the Higgs mechanism appearing in the Standard Model of elementary particles is about[1]

$$M_{EW} = 10^2 \quad [GeV]. \tag{6.3}$$

We can see that the difference between these two scales M_{Pl} and M_{EW} turns out to be considerably large. The gravity constant G is the inverse of M_{Pl}^2, and the effect is extremely small compared with the typical interaction of the Standard Model of elementary particles. Why does the nature hierarchically give us these extremely different energy scales?

The hierarchy problem appears when we consider "grand unified theories (GUT)." I have already mentioned on the GUT a little in Sect. 2.4, but here I will explain a bit more in details. The Standard Model is a non-Abelian gauge theory based on three gauge symmetries (precisely speaking, in terms of Lie groups, the symmetries are $SU(3)$, $SU(2)$, and $U(1)$). The theories which unify these three gauge symmetries into a single large gauge symmetry, while break the single symmetry spontaneously down to the symmetries of the Standard model by a Higgs mechanism, are called GUT. The research of them aims to understand in a unified way various parameters appearing in the Standard model, for example, gauge coupling constants (electric charges concerning the gauge symmetries).

There is evidence which show the existence of a grand unified theory, which is a unification of the gauge coupling constants. The coupling constants actually change a little bit depending on the energy for observation of particles with the charges. When one follows this change, the three coupling constants equal each other at a certain large energy scale. Therefore, we can expect that the three gauge symmetries are unified at this point (Fig. 6.1). This energy scale called a GUT scale and it is about

$$M_{GUT} = 10^{16} \quad [GeV]. \tag{6.4}$$

This is tremendously large compared with the energy scale of the Standard Model (6.3). This problem is called a gauge hierarchy problem.[2]

We call it hierarchy problem that either the Planck scale or the GUT scale is extremely large compared to the energy scale of Standard model. Although the

---

[1]Here, "EW" stands for electro-weak, that is, the "electro-weak theory." This electro-weak theory is the theory unifying the electromagnetism and the "weak interaction," and it is a main part of the Standard Model. In the electro-weak theory the Higgs mechanism is also used.

[2]It is considered that the scale at which the symmetry of this grand unified theory appears provides an upper limit of the energy for computing Feynman graphs of the Standard model (called "cut-off"). Since this cutoff is large as (6.4), an extremely large quantity comes out when one calculates for example a Feynman graph with loops of Higgs particles. The infinity which I mentioned previously means this cutoff in a concrete sense. Although this is made finite by the "renormalization," the value after being finite must be the energy scale of the Standard model. The gauge hierarchy problems refer to that this procedure of making things finite is not natural. This problem is also called as "fine tuning problem" or "naturalness problem."

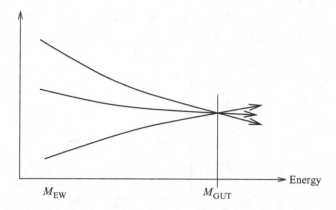

M_{EW}
M_{GUT}
Energy

Fig. 6.1 The picture of the unification of three gauge coupling constants. Depending on the energy scale of observation, the values of the gauge coupling constants change. At a certain high energy scale, the three gauge coupling constants in the Standard Model take the same value. This is considered to be the energy scale of the grand unified theory

Standard model has other problems to be clarified, we can say that the hierarchy problem is the main problem of the Standard model. Then, hereafter, I will explain that considering braneworlds can solve these hierarchy problems.

6.1.3 Braneworld

The braneworld is a concept in the elementary particle phenomenology, which appeared in the influence of the research on the D-branes of string theory. This has been developed to form a single big subject at present. The particle phenomenology is a subject which studies consequences of various phenomenological "models" supposed, in order to solve the problems of the Standard model. What we call as "models" are mainly field theory models, and the Grand unified theories as mentioned above are also a typical example. After the importance of D-branes was made clear in string theory in 1995, applications of the idea of the D-branes have been also tried in the particle phenomenology. The idea mentioned here is the concept that non-Abelian gauge theories live on the D-branes while outside the D-branes there are more extra spaces in which only gravity propagates. Although this is a conclusion of string theory, even when we do not deal with string theory some membranes on which various fields are localized are called "branes" in general. The branes are defined with this definition which is a lot milder than the D-branes in string theory.[3] As we saw in Sect. 3.1, "braneworld" means a particle

---

[3]Even in string theory the word "brane" is used often. This word is for all extended objects including fundamental strings, D-branes and such. Since the D-branes have given influence on a number of subjects, the word "brane" has various definitions.

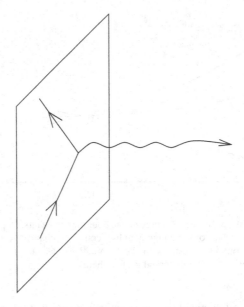

Fig. 6.2 A picture of a braneworld. Particles of the Standard Model can move only on the brane (the *straight lines* stand for their trajectories). On the other hand, gravitons can propagate outside the brane (the *wavy lines* stand for their trajectory)

phenomenology in which branes with $3 + 1$-dimensional worldvolumes are present and fields are localized on those branes in higher dimensions.

In the rest of this section, we shall see that a braneworld model can give a solution to the first hierarchy problem of the Standard Model mentioned above, that is, the problem that gravity is not included in the Standard Model, as it is unnaturally too weak. Furthermore, let us describe a sensational experimental result predicted by the braneworld (Fig. 6.2).

In 1998, N. Arkani-Hamed, S. Dimopoulos and G. Dvali considered the following situation. What if the Standard mode of elementary particles is localized on a certain $3 + 1$-dimensional brane and if only the gravity field can propagate freely in the $(3 + n) + 1$-dimensional spacetime? If gravity could propagate in such a higher-dimensional spacetime, the gravitational law would change. As (6.1) is Newton's law in $3 + 1$-dimensional spacetime, in terms of the gravitational potential V, it is rephrased as

$$V(r) = -\frac{m_1 m_2}{M_{\mathrm{Pl}}^2}\frac{1}{r}. \tag{6.5}$$

The magnitude F of gravity is given as dV/dr. If we generalize this equation for the potential to a general dimension, the generalization of the Gauss law brings it easily as

$$V(r) = -\frac{m_1 m_2}{M_{\mathrm{Pl}(4+n)}^{n+2}}\frac{1}{r^{n+1}}. \tag{6.6}$$

(Here, I introduced the Planck mass $M_{\text{Pl}(4+n)}$ in $4+n$-dimensional spacetime. When $n = 0$, it equals the original Planck mass.) In this form, this generalization has a problem: gravity controlling our solar system is based on the inverse-square law, so the power of r must not be different. So, let us consider the compactification of the Kaluza–Klein mechanism which was explained in Sect. 3.2. Let us suppose that the extra n dimensions form circles. Then, at a distance large enough compared to the radius R of this circumference, the $4 + n$-dimensional spacetime should be seen as a 4-dimensional spacetime, and the gravitational potential equation (6.6) at the $4 + n$-dimensional spacetime above should be written as

$$ V(r) = -\frac{m_1 m_2}{M_{\text{Pl}(4+n)}^{n+2} R^n} \frac{1}{r} \quad (r \gg R). \tag{6.7} $$

This is supposed to be equivalent to the inverse square law (6.5) which we observe, then we have a relation

$$ M_{\text{Pl}}^2 = M_{\text{Pl}(4+n)}^{2+n} R^n. \tag{6.8} $$

Now, our hierarchy problem is that M_{Pl} is too large compared to the energy scale M_{EW} of the Standard Model. In the present higher-dimensional theory, simply M_{Pl} is seen like this because we see eventually the gravity at a long distance scale, while the actual scale of the theory is $M_{\text{Pl}(4+n)}$. Therefore, if it is a natural theory, $M_{\text{Pl}(4+n)}$ roughly equals M_{EW}. However if we make them equal each other exactly, some gravity effects can be seen at the present accelerator experiments, which turns out to be inconsistent. So, for example, we assume $M_{\text{Pl}(4+n)} \sim 10 \times M_{\text{EW}}$, and then substitute (6.2) and (6.3) to (6.8), to obtain the following equation

$$ R \simeq 10^{\frac{30}{n}-19} \quad [m]. \tag{6.9} $$

Namely, if we like to solve the problem that gravity is too weak by the extra dimensions, this spatial size R of the extra dimensional directions is necessary.

How large is this size R of the internal space? At first in the simplest case of $n = 1$, that is, the case of only one dimension transverse to the brane, R of (6.9) almost equals the distance between the sun and the earth (about $1.5 \times 10^{11} [m]$) and so this doesn't make sense. However, if we consider the second simplest case of $n = 2$, we obtain $R \simeq 10^{-3} [m]$. Is this a sensible number? May the fifth- and the sixth- dimensional directions of the size of 1 mm exist?

In fact, in experiments, the inverse square law (6.1) of Newton has been verified only at the distance of about 1 mm for r. At a distance smaller than that for r, it might possibly deviate from the inverse square law. This will be able to be checked by experiments in the near future. In this way, the possibility that higher-dimensional spacetime exists is much closer than we expected, and so is very attractive. One millimeter is the size of letters we write in our daily life. However, all the activities such as writing and watching use interactions appearing in the Standard Model, and not in gravity. Only for the gravity, at the scale of this 1 mm, a vast extra dimensional world might be extendedly present. This elementary particle model is called a model with a large extra dimension. Here, the reason why we call "large" is that in the case of compactifications in string theory usually the radius of this compactification is the scale of Planck mass (This is about $1.6 \times 10^{-35}[m]$ in the dimension of length), while this R is much larger than that. For example, even if we take the energy scale M_{EW} of the Standard model (which is about $10^{-18}[m]$ in the dimension of length), it is still extremely larger internal space.

6.1.4 Creation of Black Holes at Experiments

The large internal space provides not only the aforementioned relation with experiments of surveying the gravity law, but also, as a matter of fact, extremely dramatical experimental results which I will describe in the following. At present, at a research facility located on the boundary of Switzerland and France, a new particle accelerator Large Hadron Collider (LHC) is at operation to find the Higgs particle, the only particle which has not been found yet in the Standard model. LHC started to operate in 2008, and as the total collision energy is scheduled to reach 1.4×10^4 [GeV], we expect to find new particles carrying the mass which amounts to the energy. A new elementary particle physics beyond the Standard model might be observed. By the way the important points of the model having these large extra dimensions is that the Planck scale $M_{Pl(4+n)}$ is near the energy scale M_{EW} of Standard model. Then it is possible that the Planck scale is an energy scale reached by the LHC. If we accelerate particles and let them collide each other with the energy reaching the Planck scale defined by the gravity, then what will happen? Quite surprisingly, a black hole will be formed!

If, at the particle collision, the distance between the particles is smaller than the Schwarzschild radius of the mass energy determined by the center-of-mass energy of the particles, black hole should be formed (see Fig. 6.3).[4] The Schwarzschild radius is the radius characterizing the size of a black hole, where objects inside the radius cannot escape. If the extra dimensions exist and one can reach the Planck scale at particles accelerators, numbers of black holes will be created there, and we may have a "black hole factory."

Once black holes are created by human beings, one might imagine that they suck up the whole earth and then human beings extinguish. However, it is not the case. This is because S. Hawking showed that small black holes should evaporate and disappear. This mysterious story of evaporation of black holes was made clear by researches on D-branes, and I will describe it in Sect. 6.3. Since the creation ratios of the species of elementary particles emitted on the evaporation of the black holes is

---

[4]Since the gravity is present in the $4 + n$ dimensions, this is a higher-dimensional black hole.

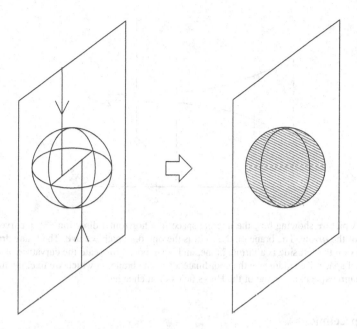

Fig. 6.3 Particles collide on the brane, and as a result a black hole is created in the $3 + n$-dimensional spacetime. In the *left* figure, particles collide inside branes (the *arrows* stand for their trajectories). When the shortest distance between the particles (called the impact parameter) is shorter than the Schwarzschild radius of the black hole in the $3 + n$-dimensional spacetime, a black holes is generated (*the right figure*)

characteristic, by observing them we can understand whether black holes are created or not at particle accelerators.

6.1.5 Solving Hierarchy Problem by Curved Extra Dimensions

In the model of the large extra dimensions introduced here, the extra dimensions with the size about 1 mm appeared. In the dimension of energy, this is about 10^{-4}[eV]. Compared with the energy scale of the Standard model, as this time it is very far from it in the opposite direction, this sounds that we don't solve the hierarchy problem after all. Then, can we construct a braneworld model with all the energy scales not so far from each other? This problem was solved by a model with a curved internal space. This model was found in 1999, and was named Randall–Sundrum model after the names of the researchers who discovered it.

For simplicity, let us suppose one extra dimension to have the whole 5-dimensional spacetime. And we consider that the spacetime is curved along the extra dimensional direction x^5. Let us take the following as a gravity field describing the

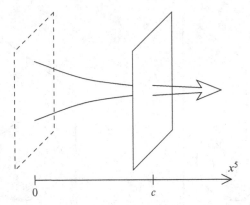

Fig. 6.4 A picture showing how the internal space (the horizontal direction x^5) in curved, by the thickness of the *arrow*. The brane on the right is the one on which we live. The brane drawn with *dotted lines* on the left side is a virtual brane, and is supposed to create the curvature of this extra dimensional space. Depending on the coordinate x^5 of the brane on which we live, the magnitude of the vacuum expectation value of the Higgs fields on it changes

curved spacetime:

$$g_{MN}(x^{\mu}, x^5) = \begin{pmatrix} -e^{-2kx^5} & 0 & 0 & 0 & 0 \\ 0 & e^{-2kx^5} & 0 & 0 & 0 \\ 0 & 0 & e^{-2kx^5} & 0 & 0 \\ 0 & 0 & 0 & e^{-2kx^5} & 0 \\ 0 & 0 & 0 & 0 & 1 \end{pmatrix}_{MN}. \tag{6.10}$$

This is a 5×5 matrix, and a metric of 5-dimensional spacetime. Among x^M ($M = 0, 1, 2, 3, 5$), the first four indices (written as x^{μ} standing for $x^0, \cdots, x^3$) show the directions along the brane (see Fig. 6.4). An interesting point of this metric is that the metric ($g_{\mu\nu}$) on the brane is multiplied by a constant number according with the location x^5 of the brane in the extra dimensional space.[5] The space represented by this metric is called an anti-deSitter spacetime (called AdS in short), and is a solution of 5-dimensional Einstein gravity with a negative cosmological constant. k is a certain constant related with this cosmological constant.

In order to see that the curved extra-dimensional space like this solves the hierarchy problem, let us consider a ϕ^4 model of $3 + 1$ dimensions localized on a brane at $x^5 = c(>0)$. As the ϕ^4 model of the $3+1$ dimensions generates a symmetry breaking and has a structure similar to the Higgs particle part of the Standard model,

---

[5] As for the physical meaning of the metric (gravitational field), see the footnote including (4.5). In the case of the present metric (6.10), the distance between two points sharing the same value of the coordinate x^5 (that is, on the same brane) is multiplied by a factor e^{-kx^5}. Even if the two points share the same x^{μ}, once their coordinates x^5 are different, the distance gets different by constants.

we can think of it as a realistic braneworld model. The action is naturally written as follows:

$$S = \int d^4x dx^5 \sqrt{-g}\, \delta(x^5 - c) \left[\frac{1}{2} g^{MN} \partial_M \phi \partial_N \phi - \frac{1}{4}\lambda(\phi^2 + m^2/\lambda)^2 \right].$$

Here, $g = \det g_{MN}$, and the Dirac's delta function[6] $\delta(x^5 - c)$ shows that the field ϕ is localized at $x^5 = c$. By a redefinition of ϕ to $e^{kc}\phi$, a little calculation shows that the following action of the $3 + 1$ dimensions follows easily (notice $\partial_5 \phi = 0$):

$$S = \int d^4x \left[\frac{1}{2}\partial_\mu \phi \partial^\mu \phi - \frac{1}{4}\lambda(\phi^2 + e^{-2kc}m^2/\lambda)^2 \right].$$

In this equation, the index $\mu = 0, 1, 2, 3$ is raised/lowered (contracted) by the flat metric in the 4-dimensional spacetime (4.7). In this action, the difference from the usual action of the ϕ^4 model (2.7) is that the location of the vacuum is

$$\phi = \pm e^{-kc}|m|/\sqrt{\lambda}, \tag{6.12}$$

which is smaller by a constant power e^{-kc}. Moreover, the important thing is that this is an exponential function.

The magnitude of the field at the vacuum is related with the energy scale of the symmetry braking. Now let us suppose that our fundamental scale is the Planck scale. Once we use this braneworld of the curved extra dimensional space, even though the original whole action of the 5-dimensions is written by the qualities of the Planck scale, on the brane at $x^5 = c$ the energy scale of the symmetry breaking becomes smaller by e^{-kc}. We want to relate the Planck scale M_{Pl} with the scale M_{EW} of the Standard model. If kc is of only the order of 50, a large hierarchy like this can be realized. In this model, the size of the extra-dimensional space is determined only by c, and so, since we think that k is determined mostly by the fundamental scale, that is, the Planck scale M_{Pl}, the extra-dimensional space is not so large to generate a new energy scale. The extremely simple assumption that the extra-dimensional space is curved turns out to solve the hierarchy problem.

What is the prediction for phenomena of elementary particles in this Randall–Sundrum model? If this model is the right one, it is expected that an interesting sign at the new accelerator will be seen. It is generation of Kaluza–Klein (KK) particles.

---

[6]The Dirac's delta function $\delta(x^5 - c)$ is the function that is infinite only at $u = 0$ while vanishes for the other values of u. The delta function satisfies the following integral equation

$$\int du f(u)\delta(u - c) = f(c) \tag{6.11}$$

for an arbitrary function $f(u)$, that is, it can extract by the integration the information of the location of the delta function.

The graviton can move also in the extra-dimensional direction, and as we look at it from the 3 + 1 dimensions, they become the KK particles with mass of $1/R$ (and its integer multiples), as we saw in Sect. 3.2. These are called KK gravitons. R is the size of the extra-dimensional space. If R is large, the mass of the KK graviton is small enough so that it might be created by the energy of the collision. Since the KK gravitons are originally a graviton, they do not have the charge of the gauge symmetries of the Standard model, and as a result of that they are not observed directly. However, in the case of the accelerator experiments, we can compare the energy before and after the collision. If we find some difference there, it means that an elementary particle without the charge, such as the KK gravitons, is generated. Namely, the KK graviton is "observed" as a "missing energy."[7]

6.1.6 String Theory and Braneworld

So far, we took D-branes in string theory in a broader sense and supposed that the "brane" equals a "subspace in higher dimensions with localized fields on it" to see that the braneworld model has a possibility to solve the hierarchy problem and predicts interesting particle accelerator experiments. There have been many trials to construct these phenomenological models from string theory. In this final part of this chapter, I will explain how to construct the Standard model by using string theory.

I explained in Sect. 5.2 that the Standard model of elementary particles has symmetries written by Lie groups SU(3), SU(2) and U(1) as a gauge symmetry, The symmetries appearing when D-branes are put on top of each other are indeed this gauge symmetry. For example, the SU(3) gauge symmetry can be realized on three D-branes put on top of each other. Then, for the case of the gauge symmetry of the Standard model, we just need to bring coincident three D-branes, two D-branes and one D-brane for each. Then how about particles other than the gauge fields? For instance, there are elementary particles carrying electric charges of both the SU(2) gauge symmetry and the U(1) gauge symmetry. If we suppose that the coincident two D-branes intersect with a single D-brane in the higher-dimensional space, then this comes up from a string connecting them (Fig. 6.5). In order for the D-branes to intersect in the higher-dimensional space, the dimension of the worldvolume of the D-brane has to be greater than four.

---

[7]In the model with large extra dimensions as we saw previously, the KK gravitons are extremely light, since the radius of the extra dimension is large. So you might wonder if a lot of the KK gravitons will be created at the accelerator. But it is not the case. These KK gravitons with the large extra dimensions have an extremely small coupling constant (given by $1/M_{Pl}$) since the interaction with particles of the Standard model is only the gravitational interaction in the 3 + 1 dimensions, so they are hardly created. On the other hand, in the case of the Randall–Sundrum model, it is known that this coupling constant is large (as about $1/M_{EW}$), because of the curved internal space. Once these KK gravitons at a particle accelerator are created and observed as the missing energy, information on the extra dimensional space should be revealed through the masses and the coupling constants.

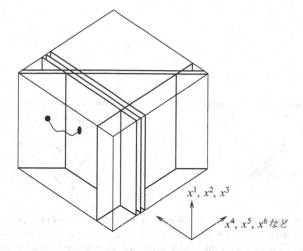

x^1, x^2, x^3

x^4, x^5, x^6 など

Fig. 6.5 The configuration of the intersection of three D-branes, two D-branes and a D-brane in the internal space. At the intersections, there are strings whose end points are on each pile of D-branes, and these generate particles with electric charges of each gauge symmetry. Since the intersection is placed inside the internal space, the dimension of the worldvolume of the D-branes must be greater than four. In this figure, the horizontal direction is the internal space, while the vertical direction is the 3 + 1 dimensional spacetime which we sense. We suppose that the internal space should be compactificated to be small

Then how do various magnitudes of the interactions among elementary particles show up? The gauge interaction and their coupling constants appear from strings at the intersection points of the D-branes as mentioned above. Then, for example, how about interactions among elementary particles appearing from the strings at different intersection points? This again has an interesting interpretation via geometry. Let us consider three distant intersections of the D-branes in the internal space (Fig. 6.6 left). The D-branes form a triangle. For the fundamental strings localized at each of the three intersections to interact, the strings must stretch for them to touch each other. Imagine the configuration, then you should find that the worldsheet of the string standing for the interaction must wrap the triangle (Fig. 6.6 Right). Since the string has a tension $1/(2\pi l_s^2)$, the probability that the worldsheet stretching to form a triangle is smaller as the area of the triangle is larger. That is, the area of the triangle formed by the D-branes in the higher-dimensional space is related to the magnitude of the interaction of the particles appearing in Standard model.[8]

Beside these, there are may other examples at which various physics occurring in the Standard model of elementary particles can be realized by a geometrical configuration of branes in higher-dimensional spaces. For example, how about

---

[8]Precisely speaking, suppose A is the area of the triangle, then the magnitude of the elementary particle interaction is $e^{-A/(2\pi l_s^2)}$. This is the probability that the worldsheet of a string with the tension $1/(2\pi l_s^2)$ extends to the area A.

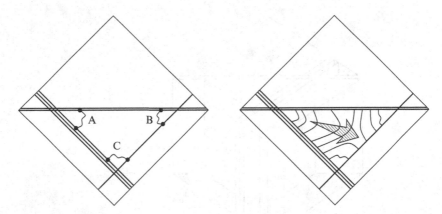

Fig. 6.6 *Left*: string A, B, C are localized at each of the intersections of D-branes (we are viewing the previous figure from the top.) *Right*: The configuration that the string A stretches and becomes string B and string C. If we follow the time-dependence, we understand that the worldsheet of the string should fill out the triangle

the Higgs mechanism? There is an idea that it is related to the reconnection of intersecting D-branes which we saw in Sect. 5.3. Once the intersecting D-branes in higher-dimensional space are reconnected, the D-branes get apart from each other, as we saw in Fig. 5.11. When D-branes get apart, the gauge symmetry is spontaneously broken, as is described in Sect. 5.2. Namely, we can see that the change of the D-brane shape in higher-dimensional space is related with the Higgs mechanism of the Standard model.

Due to the lack of room of sheets in this book I do not afford to introduce other various interesting ideas here, but you would now understand how D-branes and generalized "brane" generate new elementary particles physics, the "braneworld." D-branes offered a new paradigm called "braneworld" into the elementary particle phenomenology. On the other hand, applications of the D-branes are not limited only to the elementary particle phenomenology. They provide an extreme influence also on cosmology. In the next section, let us see that.

6.2 Application II: Inflationary Cosmology in Higher Dimensions

In the foregoing section, I introduced the braneworld model in which we brought "branes" in the higher-dimensional space and supposed that all particles except for the gravitons are localized on the branes. Since the gravity propagates in a more-than-5-dimensional spacetime there, we must change our sense of our cosmology controlled by gravity, to the one for higher dimensions. For example, in Sect. 5.2, monopoles are represented by a configuration of D-branes in higher dimensions, and I explained application of the interpretation, then how about cosmology? This is also an important problem of string theory: how can we include cosmologies in the framework of string theory?

String theory is a framework which can handle gravity in a way of quantum theory, and stringy effects appear at the energy scale of strings, that is, at the energy of the order of $1/l_s$. As a result of that, cosmological physics at high-energy region should be closely related with string theory. A typical one of them could be the origin of the universe, namely, the big bang. The big bang is a singularity in terms of the gravity theory, so one can not describe what kind of physics occurs at that point (time) by the general relativity. This is the same situation as the singularity of black holes. I will explain, in the last of this section, a little about what kind of possibility string theory gives for the big bang. In this section, I mainly explain cosmological inflation for which there are interesting research results of relations between cosmologies and D-branes.

The universe is expanding at present, and it is considered that there exists a period of a very fast expansion in the early stage of the universe. This is called inflation cosmologies. This inflation is accompanied with an interesting higher-dimensional interpretation with D-branes: the inflation cosmologies can be naturally explained by D-branes moving in higher-dimensional spacetime.

6.2.1   Inflation Cosmology

At first I will have a brief explanation of necessity of the inflation cosmologies, and also the mechanism. The inflation cosmology is a conjecture that there was a period with an exponentially fast expansion of the universe at its early stage. A. Guth and K. Sato independently proposed in 1981 the model as the one solving various "problems" which appear in the evolutional process of the universe beginning with the big bang. The problems are mainly the following three:

- Monopole problem:
  Since the temperature was extremely high in the early universe, there was a period in which the energy density of the universe is higher than that of the spontaneous symmetry breaking of grand unified theories. As the universe got colder and colder, the spontaneous symmetry braking occured, and a number of monopoles should have been generated, associated with that (the Kibble mechanism). However, we have not observed the monopoles yet. This can be explained as that the density of monopoles was diluted in the period of the inflation when the universe expanded rapidly.

- Horizon problem:
  There is an observation that cosmological background radiation is homogeneous at any direction of the universe seen from the earth. This is a problem in the sense that it appears to relate regions of the universe which are causally disconnected. Because there is no information communication which runs beyond the speed of light, two regions which are too far from each other for reaching by the speed

of light are out of the "horizon" and they cannot be related with each other. This problem is resolved by the inflation as follows: before the inflation the two regions were casually connected with each other actually, but after the inflation they are isolated by a distance which has no casual contact.

- Flatness problem:
  The expansion of the universe is described by the Einstein's generally relativity. If in the universe the amount of matters (precisely speaking, the energy density) were too large, the universe should have contracted soon and would have not expanded to the present extent. (The universe like this is called a closed universe.) On the other hand, if the amount of matters were too small, the universe would have expanded so fast that there would have been no time for the structure of the universe such as galaxies to form. (This universe is called an open universe.) The present energy density of the universe sits just in the middle of these, and is called a plain universe. But the standard Big bang model of the universe cannot explain why this is the case. If the inflation exists at the early stage of the universe, by the rapid expansion of it makes the curved space stretch to become flat, then this problem is solved.

Then, how the exponential expansion of the universe can be realized? It can be realized (though I do not explain it in detail) by introducing a positive cosmological constant in the action of general relativity. The cosmological constant is a potential energy of vacuum. However, the inflation must end after the universe expanded exponentially to a certain extent. From the first place, the present universe does not expand exponentially. Then, it leads to that the potential energy of the vacuum has to vary in time. Though at first the potential energy is positive, it has to become zero after a while.

In the standard inflation cosmology, this framework is supplied by introducing a field $\Phi(x^\mu)$ called an inflaton. Let us see Fig. 6.7. At a certain time, the inflaton is at the top of its potential, and its value changes in time. And finally the inflaton field reaches the bottom of the potential and the potential energy becomes about zero, and the inflation is over.

In string theory, through the use of D-branes for braneworlds, this inflation can be realized in a very interesting way, geometrically and higher-dimensionally. The inflation is interpreted as a motion of the D-brane in a higher-dimensional spacetime, and the end of the inflation is described by a collision and a pair-annihilation of the D-brane and an anti-D-brane. Let us see this "brane inflation" in the following.

6.2.2 The Motion of D-Branes and Inflation

When D-branes exist, on them there are schalar fields describing the position of the D-branes in the direction transverse to the worldvolume (Sect. 4.2). Those are

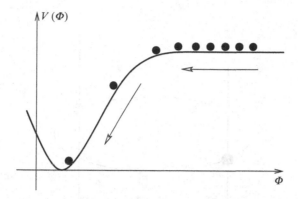

Fig. 6.7 A time dependence of the value (*black circle*) of the inflaton field Φ, which is used in inflation cosmologies. At first when the potential energy is positive, Φ moves slowly. After a while, the field slips down the potential and reaches zero

a natural candidate for the inflaton field Φ which is a scalar field. A concern is the potential of the inflaton field. How can we provide the potential as in Fig. 6.7 for D-branes in string theory?

Suppose we set some other D-brane, in addition to the D-brane we considered. As far as they are placed parallelly, as I mentioned in Sect. 5.1, there is no force between the D-branes. That means that the scalar field Φ has no potential. As the distance between the D-branes can be arbitrary, any Φ is a stable solution. (Precisely speaking, this Φ is given by the D-brane separation $\Phi^{(1,1)} - \Phi^{(2,2)}$ with the scalar field on each D-brane $\Phi^{(1,1)}$ and $\Phi^{(2,2)}$. See Sect. 5.2.) However, let us replace the other D-brane by an anti-D-brane. The anti-D-branes has the Ramond–Ramond charge of the sign opposite to that of the D-brane. For two D-branes, gravity and the Ramond–Ramond force cancel with each other, while this time they sum up and the D-brane and the anti-D-brane attract each other. We can interpret the attraction as caused by a generated potential of the field Φ.

In order to generate enough inflation, it is known that the upper part of the potential of the inflaton field has to be flat as in Fig. 6.7. In order to attain this flatness by D-branes, let us consider the following set up. For simplicity, we suppose that in the compactified space, the direction along which the D-brane and the anti-D-branes is separated is a one-dimensional circumference. And let us suppose that the D-brane and the anti-D-branes are placed on the anti-podal points (Fig. 6.8). Since when they are at the anti-podal points the forces from the both sides are in balance and they do not move, this state is "mearginally stable." However, if they are shifted even a little from that position, their balance is broken and they change their positions rapidly. Therefore, around the anti-podal points the potential becomes flat enough, and the inflation is generated. It is very interesting that the inflaton field has a higher-dimensional geometric interpretation, in this way.

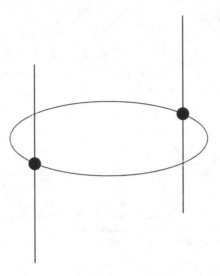

Fig. 6.8 A configuration of D-branes which generates an inflation. When the internal space is a circumference and the D-brane and the anti-D-brane are placed on the anti-podal place, this is mearginally stable. If they are shifted even a little, they start to move closer. The distance between the D-brane and the anti-D-branes becomes the inflaton field

6.2.3   The End of Inflation and Annihilation of D-Branes

Then, as the D-brane approaches the anti-D-brane closer, what happens if they collide with each other at last? They should pair-annihilate. Namely, the end of the inflation is a pair annihilation of D-branes!

The pair annihilation of D-branes was described in Sect. 5.3. And the important thing there is that a tachyon appears in the oscillation modes of the strings $\langle 1, 2 \rangle$ and $\langle 1, 2 \rangle$ connecting the D-brane and the anti-D-branes, and the tachyon vacuum-condenses. In fact, this tachyon does not appear if the D-brane and the anti-D-brane are far enough from each other. Let us see why this is the case. As we saw in (5.5), when a string is stretched between D-branes, the length of the string amounts to the distance between the D-branes, which contributes to the energy of the string. In view of the fact that the second term of (5.5) contributes to the mass squared of the tachyon, the tachyon mass coming from the string between the D-brane and the anti-D-brane is

$$ m^2 = -\frac{1}{l_\mathrm{s}^2} + \left(\varPhi^{\langle 1,1 \rangle} - \varPhi^{\langle 2,2 \rangle} \right)^2 . \tag{6.13} $$

Here we have used that the distance between the D-brane and the anti-D-brane is given by $2\pi l_\mathrm{s}^2 | \varPhi^{\langle 1,1 \rangle} - \varPhi^{\langle 2,2 \rangle} |$. The factor $2\pi l_\mathrm{s}^2$ is deduced in string theory so that the value of the field is translated to fit the dimension of distances. This formula tells that m^2 is negative only when the D-brane and the anti-D-brane are close

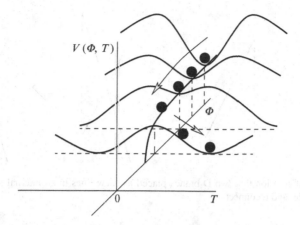

Fig. 6.9 The potential of the inflation by the D-brane and the anti-D-brane. The value of the fields (*blob*) moves slowly in the space of the inflaton field Φ and the tachyon field T

enough. The tachyon does not appear when the distance is larger than a certain value. However, when the distance is small enough, the tachyon with a negative mass squared appears, and the D-brane and the anti-D-brane pair-annihilate via the vacuum condensation.

The relation between the two kinds of fields, the tachyon field T (in fact, there are two tachyon fields, coming from strings $\langle 1, 2 \rangle$ and $\langle 2, 1 \rangle$) and the scalar field $\Phi = \Phi^{(1,1)} - \Phi^{(2,2)}$, is described, and now the picture of the total potential is shown in Fig. 6.9. The black blob stands for the value of the fields (T, Φ), and the figure shows how it moves as time goes. At first, since the mass squared of the field T is positive, it remains at $T = 0$. The field Φ moves slowly in the slowly varying potential in the direction to smaller values of $|\Phi|$, during which the inflation takes place. When $|\Phi|$ decreases small enough, the mass squared of the field T becomes negative, then the tachyon condensation occurs. In Fig. 6.9, the direction of the motion of the field changes to the right, and the inflation ends when it reaches the bottom of the potential.

This mechanism, in which a scalar field (tachyon T in our case) is introduced in addition to the field creating the inflation (the inflaton field Φ) so that the inflation naturally ends, is called a hybrid inflation. This is a model often used in inflation cosmologies. The string-theoretical description of the inflation by a pair annihilation of a D-brane and an anti-D-brane realizes the hybrid inflation quite naturally.

An interesting physics appears when the D-brane and the anti-D-brane pair-annihilate. As we saw in Sect. 5.3, in the pair-annihilation process of the D-branes, vortex solitons are generated in the condensation of the complex tachyon field (Sen's second conjecture), and these are D1-branes in the present case. Namely, there is a possibility that string-like objects extending like one-dimensional strings in space may be generated at the end of inflation. These resemble the cosmic strings we saw

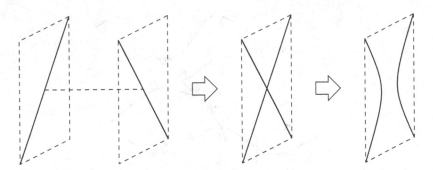

Fig. 6.10 A configuration that two D-branes placed as skew lines in an internal space approach each other, collide, and reconnect

in Sect. 3.1. In fact, recently a possibility was discussed that the D1-branes may be realized as the cosmic strings.[9] Once the cosmic strings are actually observed, they might be the D1-branes! That is, we might be able to see the huge D-branes extending in the sky. What a great and interesting possibility it is!

Once the inflation comes to end by the pair-annihilation of the D-brane and the anti-D-brane, then where is our brane representing our world? Although this problem can be solved by supposing that there are originally third D-brane in addition, let us introduce an idea using reconnection of intersecting D-branes, here. Let us suppose that two D-branes are placed at skew lines in an internal space (Fig. 6.10). Since these are a little bit deviated from the parallel state, an attraction between them is generated. If we choose the angle of the skew lines to be π, this is a D-brane and an anti-D-brane, then all the D-branes vanish. However, this

---

[9]Once cosmic strings are observed, how do we distinguish the vortex solitons of a field theory in Sect. 2.3 from the D-branes? This is an interesting question. Concretely speaking, it has been argued that there may be a difference in the number of the cosmic strings observed, for the field theory vortex solitons and for the D-branes. Figure 3.2 shows how the cosmic strings collide and reconnect with each other, and then form closed loops which contract to vanish. In this way, as the reconnection is a reason for the number of cosmic strings to decrease, what kinds of situation the reconnection occurs relates to how many cosmic strings remain in the end. It is known that the reconnection is classical in the case of the vortex solitons of the field theory, while it is probabilistic in the case of the D-branes, in fact. This difference should result in the number of the cosmic-string-like objects which we observe, at the end. By the way, we know that the D1-branes are vortex solitons coming from the tachyons of the D3-brane and the anti-D3-brane, then what is the origin of this difference? The reason is that in the annihilation of the D-branes there are related other stringy modes (infinite massive modes) which we omit, in addition to the tachyon. From the string connecting a D3-brane and an anti-D3-brane, there appear many oscillation modes as well as the tachyon, and these infinite number of modes condense too at the same time as the tachyon condensation. This is the reason why the D1-branes are different from just the vortex solitons. In the discussion of the annihilation of D-branes in Sect. 5.3, I described this points briefly. To prove the Sen's conjecture was difficult (however finally it was proved), because of the difficulty to handle these infinite number of modes. The theory to handle the infinite number of modes simultaneously is the string field theory which appeared in Sect. 4.1.

time it is not the case. As we saw in Sect. 5.3, the tachyon field is localized near the intersection point, and the tachyon condensation bring the reconnection of the D3-branes. Therefore, even if the inflation ends, in this case the D-branes do not disappear, and they remain almost parallel to each other.

6.2.4 Why Do We Live in Four Spacetime Dimensions?

So far, I have described the geometrical method to realize the inflationary cosmology by using D-branes. The natural motion of D-branes that a D-brane approaches an anti-D-branes and they pair-annihilate explains the inflation. Now, the biggest mystery in cosmology is the question of why our world is $3 + 1$ dimension. For this mystery, the pair-annihilation of D-branes gives us a hint. It is called a branc gas cosmology, which is one of the subjects under active research.

If all the elementary particles and the interactions are described by superstring theory, they are in the 10-dimensional spacetime. Many strings and D-branes move in the 10-dimensional spacetime. So, let us suppose that, as in the beginning of our universe in the big bang cosmology, this 10-dimensional spacetime may have been extremely hot at the beginning. Then, because of this thermal energy, D-branes and anti-D-branes should have pair-created one after another. However, when the 10-dimensional spacetime gradually cools down, these D-branes and anti-D-branes disappear by pair-annihilation. When they disappear, as described in the Sen's second conjecture, lower-dimensional D-branes and anti-D-branes are created. These generated lower-dimensional D-branes and anti-D-branes form pairs and collide and pair-annihilate further, but you may recognize that the process of this "pairing" is more difficult as the dimension of the D-branes is lower. D7-branes and anti-D7-branes in 10-dimensional spacetime intersect easily and pair-annihilate, except for some special situation. This is also the case for D5-branes. However, in the case of D3-branes, since the dimension of their worldvolume is too small compared to the 10-dimensional spacetime, it takes very long time to form pairs. By this reason, we argue that our universe becomes D3-branes, that is, of $3 + 1$ dimensions. It is very interesting that this simple mechanism may determine the dimension of our universe.

You might admit that the brane gas cosmology also gives quite an interesting interpretation for the big bang, namely, the beginning of our universe The D3-branes which are our universe started with a pair-annihilation of the D5-branes and the anti-D5-branes. Therefore we can interpret that the big bang is a pair-annihilation with a collision of higher-dimensional D-branes and a creation of D3-branes from there.

Since the gravity can freely propagate also in the internal space, we must explain why the internal space becomes round small, for this scenario to be right. There have been various trials for solving this problem, for example by wrapping D-branes along the direction of the compactified internal space. The big mystery of our universe, why we live in the 3-dimensional space, might be solved by dealing with geometrical objects in higher-dimensional spacetime, by using string theory.

6.3   Application III: Black Hole Entropy

Black holes are solutions of equations of motion of general relativity, and one can say that they are the most important objects predicted by the gravity theory. As for their observation, for example it is believed that there exists a huge black hole at the center of the Milky Way Galaxy where our solar system resides, but speaking exactly, it has not been observed yet, because it is a "black hole." What has been observed actually is various interesting physical phenomena expected near black holes.[10]

However, the black holes are not only interesting objects in observation but also important objects theoretically. I mentioned in Chap. 6 that the problem of the standard model of elementary particles is that it does not contain gravity. The reason for this is that there is a big problem that gravitational field theory cannot be quantized. The gravitational interaction is "accidentally" small enough compared to the other interactions appearing in the standard model of elementary particles, and because of this the standard model succeeds even though gravity is ignored. However, black holes make space time extremely curved, so it is considered that around it gravitational effects are quite large. If string theory is a quantum field theory unifying all the interactions including gravity, we should be able to describe precisely what happens at the black hole. What kinds of knowledge does string theory give us about black holes?

In this section, first, we see a problem called "information loss" hidden among the black holes and quantum mechanics, and take an overlook at "black hole thermodynamics" related with it. The thermodynamics of black holes is thermodynamic formulas which black holes appear to follow. One of the relations is a formula for "black hole entropy," called Beckenstein–Hawking formula. In the standard thermodynamics, there is a derivation of the relations from the viewpoint of statistical mechanics, since a number of microscopic particles gather and behave statistically, and the thermodynamics is understood as macroscopic relations followed by the whole bunch of particles.[11] However, in the case of the black holes, this viewpoint of the statistical mechanics is completely lost. This is because the thermodynamics of black holes came from just an analog, as I will describe in the following.

What are the microscopic constituent elements reproducing the thermodynamics of black holes? After all, what are black holes made of? D-branes show up here to answer this question. As we saw in Sect. 4.2, some kinds of black holes are identified

---

[10]For instance, since black holes accelerate objects around them and absorb all of stuffs, there appears a disk of matters rounding at a high speed around the black hole, which is called an accretion disk. The acceleration energy is emitted as strong X-rays. We observe these X rays and guess the mass and the volume of the stars, the would-be black holes.

[11]In this section, basic knowledge of thermodynamics is supposed. The entropy S is a quantity characterizing "disorder" of the whole system where many microscopic elements interact with each other. When the microscopic state number of the system is d, the entropy is written as $S = k_B \log d$ (here, k_B is the Boltzmann constant).

as D-branes. And D-branes are defined by strings, the microscopic constituent elements. According with this idea, counting of the number of states of black holes was made by using D-branes, and their entropy was derived from the viewpoint of the statistical mechanics in the microscopic way. The formula of Beckenstein and Hawking was derived exactly. This proof was sensational, as it became an important example showing that string theory actually works as a quantum gravity theory. At the end of this section, I introduce how this black hole entropy was derived.

6.3.1 Black Holes and Quantum Mechanics

The big problem when we deal with black holes quantum mechanically is the information loss problem. This is the problem that although in quantum mechanics time evolution does not lose any information black hole exists and then informations are lost and contradict quantum mechanics. Before explaining this problem, we need to see the thermodynamics of black holes first.

In Sect. 4.1, the radius of a sphere which is an event horizon of a black hole, the Schwarzschild radius, is given by $r = 2Gm$. (Here, G is the gravity constant, and m is the mass of the black hole.) Once any particle falls into the event horizon of the black hole, it cannot escape from there ever. Therefore the mass of the black hole only increases. According with it, the radius of the black hole grows bigger, and the area of the event horizon increases. This resembles the second law of thermodynamics, which states that entropy S never decreases.

Actually, the following existence of the thermodynamic low of black holes was pointed out. The first low of thermodynamics is written as $dE = TdS$ (here E is the total energy and T is the temperature), and the second low of thermodynamics states that entropy never decrease, $dS \geq 0$. For black holes, the entropy is replaced by the area A of the event horizon. And the energy E is replaced by the mass of the black hole, and the temperature T is by the Hawking temperature. This is called a thermodynamics of black holes.

Hawking temperature is the temperature of black hole, which was suggested by Hawking. Let me explain why black holes have the temperature. Objects carrying temperature have radiation corresponding to the temperature. Since black holes only swallow particles and do not spew them out, one may think that black holes do not radiate. However, this is just a classical picture, and if we consider quantum mechanics near the event horizon of the black hole, it turns out to radiate, as follows. Let us suppose that a particle and an anti-particle are pair-created just outside the event horizon (Fig. 6.11). Due to quantum mechanics, there are pair creations everywhere. And let us also suppose that the anti-particle is eventually absorbed by the black hole while the particle safely escape outside. Then, this process looks as if the particle is emitted from the black hole as seen from far away, and the black hole radiates. This is called Hawking radiation. Since the radiation has a spectrum characterized by the temperature of the radiator, the temperature of the black hole is determined by that. This is called Hawking temperature, which is given as follows

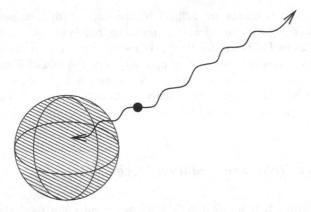

Fig. 6.11 A configuration of the Hawking radiation. A particle and an anti-particle are pair-created outside the event horizon of a black hole, and one is absorbed by the black hole while the other "runs away" from the black hole. The *big ball* stands for the event horizon of the black hole, and the *black blob* means the pair creation, and the *wave lines* stand for the trajectory of the particles

for the case of Schwarzschild black hole:

$$T = \frac{1}{8\pi Gm}. \tag{6.14}$$

Here I used the unit system that the Boltzmann constant k, the speed of light c, and the Planck constant $h/2\pi$ all set to 1. Using this formula, with the first low of thermodynamics, we can express the entropy precisely by the area A of the event horizon of the black hole:

$$S = \frac{1}{4G}A. \tag{6.15}$$

This is called Beckenstein–Hawking formula. Although here I described the story only for the Schwarzschild black holes, it is known that this formula is applicable for more general framework, such as charged black holes and black holes in higher dimensions.

Once the black hole emits particles by the radiation, then the mass of the black hole decreases. This is called evaporation of black holes, and in the process of the evaporation, the problem of the information loss in quantum mechanics shows up. The black hole itself was formed after some matter which was originally there collapses gravitationally and shrinks. The original matter should have been described by quantum mechanics, and we can learn the quantum mechanical state of it (this is called a pure state). However, once a black hole is formed after the gravitational collapse, the evaporation process is caused by the quantum-mechanical pair correlation of particles outside the black hole as mentioned above, and because the particles came out by the evaporation are completely randomly chosen, they do not have the original information of the matter which forms the black hole. The state like this is called a mixed state. In this way, the process of the formation and the evaporation of black holes loses the original information. This is called information paradox problem of black holes.

Then, can string theory, which has been told that it is a quantum gravity theory for long years, solve the information paradox problem? This problem is considered to relate with the question of what happens at the end of the black hole evaporation. The development in string theory which I will explain next has not yet been able to solve this big problem, but offers an answer to the following question as a single step toward the solution of the information paradox problem: What is the microscopic states of black holes? And the answer is "states of strings on D-branes."

In the black hole thermodynamics, it is just that the thermodynamic formula can be applied to black holes, that has not been calculated from some statistical system of microscopic states. However, string theory provides what this statistical system is, and for a certain class of black holes, a calculation of entropy from this statistical system can be made, and it results in the entropy formula of Beckenstein–Hawking.

6.3.2 D-Brane Configuration Representing Black Holes

In Sects. 4.1 and 4.2, it became clear to us that BPS black holes with electric charges can be identified as D-branes in string theory. Surprising in this identification is that D-branes are originally unrelated with curved spacetimes while are defined as a space on which strings can end. Therefore, in order to examine characteristics of black holes, we can use the D-brane, a simple space.

The entropy of a black hole should be determined by counting the microscopic states of the black hole in statistical mechanics. Once D-branes are identified with black holes, it should be clear what these microscopic states are. They are the states of a string which has end points on the D-branes. In 1996, A. Strominger and C. Vafa introduced the entropy of a black hole with electric charges in five-dimensional spacetime, by counting the states of the string. Here, we shall take the example given by C. G. Callan and J. Maldacena which is easier to understand.

The black hole we consider is a black hole in five-dimensional spacetime. Although we may deal with four-dimensional spacetime which is familiar to us, we shall consider a five-dimensional one for a simplification of the calculations. String theory is defined in ten-dimensional spacetime, and we will make a Kaluza–Klein compactification (see Sect. 3.2) of the five-dimensional part $(x^5, x^6, x^7, x^8, x^9)$ to a circumference for each. Then, at low energy, this becomes a supergravity theory in five-dimensional spacetime spanned by $(x^0, x^1, x^2, x^3, x^4)$. As well as the gravity field and the Ramond–Ramond gauge field, there are gauge fields coming from the Kaluza–Klein mechanism. As a solution of this supergravity theory, we consider a black hole which keeps supersymmetries and has electric charges as follows:

- Electric charge Q_1 of the Ramond–Ramond field C_{MN}.
- Magnetic charge Q_5 of the Ramond–Ramond field C_{MN}.
- Electric charge N of the Kaluza–Klein gauge field.

Here the electric charges of the Ramond–Ramond field is that of $C_{5\mu}$ ($\mu = 0, 1, 2, 3, 4$). Moreover, the magnetic charge is a source for the magnetic field $H_{\mu\nu\rho}$, where $H_{MNR} = \partial_M C_{NR} + \partial_N C_{RM} + \partial_R C_{MN}$ is the field strength made by C_{MN} (see (2.1) for the field strength in electromagnetism. Remember the Hodge dual given in Sect. 4.1). And the last electric charge N is that of the Kaluza–Klein gauge field which is $g_{5\mu}$. The reason why we introduce many charges like this which looks cumbersome is in fact that easier charged BPS black hole solutions have problems that their event horizons coincide with their singularities. On the other hand if one introduces various electric charges like this example, the event horizon sets apart from the singularity and has a nonzero area. When one substitute this area to the Beckenstein–Hawking formula and calculate the entropy, one can get the result

$$S = 2\pi \sqrt{N Q_1 Q_5}. \tag{6.16}$$

As we will see below, this result is reproduced by D-branes, including its numerical coefficients.

6.3.3   Derivation of Entropy Formula by D-Branes

First, let us see what kind of D-branes corresponds to this black hole with the electric charges and the magnetic charge. You may understand that the electric charge of the Ramond–Ramond field can be created by D1-branes wrapping by Q_1 times the direction of x^5 among the internal space. Next, in the same way, the magnetic charge of the Ramond–Ramond field can be produced if D5-branes wrap by Q_5 times all the internal space (x^5, x^6, x^7, x^8, x^9). And as for the last electric charge of the Kaluza–Klein gauge field, the field was originally a gravity field $g_{5\mu}$. Since electric charges of gravity field stand for momenta, the electric charge N of $g_{5\mu}$ corresponds to a string moving with momentum N in the direction of x^5. A picture showing the configuration of the D-branes and the strings is Fig. 6.12. When we show complicated configurations of D-branes like this, we often use the following table representing the directions along which the D-branes extend.

	x^0	x^1	x^2	x^3	x^4	x^5	x^6	x^7	x^8	x^9
D1	○					○				
D5	○					○	○	○	○	○
string	○					→				

Here, the arrow "→" stands for strings moving in that direction.

As it is expected that the number of states of the black hole is that of the states of the string connecting these D-branes, let us consider what kinds of strings exist concerning the D-brane configuration. Since there are Q_1 D1-branes and Q_5 D5-branes, we have the following species of strings:

- D1-D5 string: connecting the D1-branes and the D5-branes. There are $Q_1 Q_5$ species.

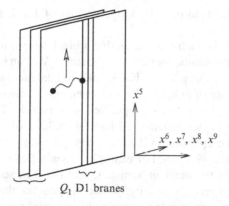

Fig. 6.12 A configuration of D-branes corresponding to the five-dimensional black hole with the electric and the magnetic charges. All D-branes wrap the internal space. Strings connecting the D1-branes and the D5-branes move with momenta in the direction of the internal x^5

- D1-D1 string: connecting the D1-branes. There are $(Q_1)^2$ species.
- D5-D5 strings: connecting the D5-branes. There are $(Q_5)^2$ species.

Let us excite these strings and let them have the momentum N. At this stage, if many of the first D1-D5 strings are excited, it is expected that the second and the third strings acquire masses and are not excited. The reason is as follows. For instance, as we saw in Sect. 5.2, once parallel D-branes are set away from each other, strings connecting them acquire masses. For the D-brane to change its position to make itself separate from the other, the scalar field representing the location of the D-brane needs to change its vacuum expectation value. This change can be interpreted as a collective motion of many elementary excitations. This is the same relationship as that between solitons and elementary excitations (see Sect. 2.2). In the present example, when many strings connecting the D1-branes and the D5-branes are excited, the other kinds of strings acquire their masses, and they are not excited.

As a result, only a lot of D1–D5 strings are excited, with the internal momentum N. The number of states $d(N)$ of strings with the total internal momentum N is, when N is large, known to be given as

$$d(N) \sim e^{2\pi \sqrt{cN/6}}. \qquad (6.17)$$

(As the derivation of this concerns quantization of a string, I shall not describe it here. Basically, it is a counting of the number of possible oscillation modes of the string.) Here the constant c is the quantity related with the numbers of the kinds of the strings, and is given by $c = 6Q_1 Q_5$ corresponding to the D1-D5 strings we consider. Using this, by taking a logarithm, we derive the entropy as

$$S = \log d(N) \sim 2\pi \sqrt{N Q_1 Q_5}. \qquad (6.18)$$

This is indeed the Beckenstein–Hawking formula of black hole entropy given by (6.16).

In this way, it was shown by D-branes that a black hole entropy can be derived actually by counting the number of microscopic states. We can say that string theory succeed in taking the first step to link black holes to quantum mechanics.

However, the physics of D-branes has not been able to answer the big question of what remains after black holes evaporate. As a matter of fact, the Hawking temperature T of the higher-dimensional BPS black hole which we considered is zero, thus it does not evaporate. On the other hand, the temperature of evaporating black holes is of course nonzero. For example, when a Schwarzschild black hole evaporates, in the end the mass m vanishes, and one can see in (6.14) that the Hawking temperature reaches infinity. What happens at the last moment? To solve the information loss paradox completely, we might need further study on D-branes and string theory.

In the next section, I will introduce more intriguing and new duality born out of the identification of black holes and D-branes. There it is revealed that a certain gravity theory in five-dimensional spacetime and a non-Abelian gauge theory in four-dimensional spacetime are in fact equivalent to each other. Two physical systems which have different dimensions and have different kinds of theories turn out to be related beautifully by D-branes. Furthermore, this duality has a possibility of resolving the difficulty in QCD which was described in Sect. 2.4.

6.4 Application IV: Holography – Quarks and Strings

As was described in Sect. 2.4, Quantum Chromodynamics (QCD) has calculational difficulties. Since in QCD the gauge coupling constant is large, as a matter of fact one cannot calculate scattering of various particles in perturbation theory which is in an expansion in powers of the coupling constant. In fact, in QCD, the issue is not only the calculational difficulty, but also a more fundamental one which should be solved. It is the problem called "quark confinement."

QCD is a non-Abelian gauge theory based on a gauge symmetry SU(3), and is contained in the standard model of elementary particles. There are eight gauge fields corresponding to the symmetry, and they are called gluons. And elementary particles carrying the electric charges of this symmetry are called quarks. For example, proton consists of three quarks. However, we have never observed a single quark by itself in any experiment. Why is it?

Heuristically, one may speculate that, since the coupling constant of QCD is too large, quarks bind together so that any single quark may not be able to exist alone. However, it has never been proved theoretically. After all, at present we have not understood yet QCD, the theory with a large coupling constant in which quantum theoretical effects (many loops in Feynman graphs as in Fig. 2.18) are essential. This is really a serious problem. Actually, one of big seven mathematical millennium problems proposed by Clay mathematical institute in the United States is concerning this QCD, and one million dollars will be awarded for a solution of the problem related with this.

D-branes give us quite an innovative idea, as an approach to this problem. It is called "holography" which is introduced in this chapter. In Sect. 5.2, we saw that when D-branes are on top of each other, a non-Abelian gauge theory appears on them at low energy. And in Sect. 4.2, we saw that D-branes can be identified with black holes. As a combination of these, the non-Abelian gauge theory can be related with the black holes, a gravity theory with a curved spacetime. Thanks to this new idea, we can calculate various properties of theories with a large coupling constant such as QCD by using gravity theories! Non-Abelian gauge theories and gravity theories become equivalent to each other – this is a duality, a corresponding principle.

In this duality, the spacetime dimension defined by the non-Abelian gauge theory, which is $3+1$ dimensions, is different from that defined by the corresponding gravity theory. This is related with the fact that the gravity theory came from the black holes and they are higher-dimensional black branes appearing in string theory. The case in which physical theories with different dimensions are equivalent to each other is called holography.

First of all, how does this correspondence appear? And how does the theories with different dimensions correspond to each other? And, what kinds of new calculational method may be given by this correspondence to theories with a large coupling constant such as QCD? In this section, let me explain this surprising duality.

6.4.1 String Theory and Quantum Chromodynamics

Historically, string theory appeared as a theory describing hadrons. I would like to introduce how to relate strings with the hadrons.

Hadrons stand for bound states constituting of two or three quarks in QCD in the standard model, and protons and neutrons are included among them. It is known that, as a result of experimental observation related to hadrons, when one consider groups of hadrons sharing the same quantum numbers (conserved numbers proper to particles such as electric charges) except for the spin J, the mass squared of these particles is a linear function of J, and the linear coefficient of J is shared by all the groups. This is called Regge trajectory.

The Regge trajectory can be explained by an assumption that hadrons are made of something like a string. Let us remember the spectrum in the open string theory, that is the mass formula (3.7) which we saw in Sect. 3.2. The mass squared m^2 is proportional to an integer-multiple of the string tension, and the integer is related to the spin, in the manner that the excitation is a tachyon namely a scalar field (whose spin zero) when the integer is zero, and it is a massless gauge field with spin 1 when the integer is 1. This fact follows also for larger integers and larger masses, and actually it becomes the Regge trajectory.

We can understand that this reproduction of the Regge trajectory by the string originates in the fact that the string has a constant tension. Let me make this reason clear. It is known that the Regge trajectory is satisfied in the following general cases. Suppose that there are two particles and they form a bound state via a force between them. We suppose that the potential standing for the force is proportional to the distance L between the particles.

$$V(r) \propto L. \tag{6.19}$$

When there is a potential like this, as longer distance costs more energy, the two particles attract each other with a strong force so that they are expected not to be observed independently. This potential is called a confining potential. When two particles rotate around each other with this attractive force, the total energy (namely, the mass of a "hadron" if we regard the system as a single particle hadron) is known to be proportional to the internal angular momentum coming from the rotation. Therefore, the confining potential turns out to reproduce the Regge trajectory.

Let us see that this confining potential is the very string. For that, we had better think why the potential like this appears, first of all. In electromagnetism, the form of the potential is $1/L$, because the electric flux (the electric field) comes out homogeneously from an electron in the three-dimensional space. Now, let us suppose that this is not homogeneous but it extends only in the direction to the particle which is paired (Fig. 6.13). Since the total electric flux should be conserved, only on this line the electric flux has the total value of the conserved constant. That is, the electric flux is constant on the line. Then, the electrostatic potential turns out to be (6.19). This is because the electric flux is constant on the line and it is given by a derivative of the potential. Therefore, the potential like (6.19) describe an electric flux which is localized linearly and connects particles as a straight line. Strings are identified with the line electric flux, in the sense that strings have a constant energy per unit length (tension). Namely, we may come to a picture that a string connect two particles and they rotate, forming a hadron. The confining potential and the resultant Regge trajectory can be derived if one suppose that strings connect particles.

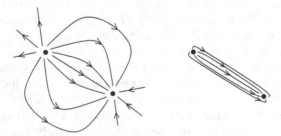

Fig. 6.13 *Left*: A configuration of an electric field between an electron and a positron in electromagnetism. *Right*: The existence of the confining potential corresponds to some string consisting of total electric fluxes concentrating on a line

In QCD, the particles making pairs like this are considered to be quarks. Then, the interaction corresponding to the strings should be the gauge interaction by gluons. However, the reason why the gauge interaction of the gluon has the confining potential has not been proved. This is indeed a problem of QCD.

QCD has this unsolved mystery, but historically string theory had a more serious problem. It is a problem of the spacetime dimensions. String theory is well-described consistently only in 26 dimensions in the case of bosonic strings, but then it cannot describe hadrons in four-dimensional spacetime. On the other hand, why was QCD established as an important part of the standard model in spite of the unsolved mystery? The reason is that a property called "asymptotic freedom" has been shown in calculations of a perturbation theory of QCD. In the late 1960s, experimental results of deep inelastic scatterings at which electrons collide with hadrons can be well reproduced by the quark model, once in the hadrons quarks are supposed to move freely almost at the speed of light. However, why don't quarks fly out of the hadrons? In 1973, D. Gross and F. Wilcek, and D. Politzer found[12] the "asymptotic freedom" which means that, once the quantum effect of the coupling constant is implemented in QCD, at short distance the coupling constant becomes small while at long distance it becomes large. The term "asymptotic" means the asymptotically high energy region, since in terms of energy, the short distance corresponds to the high energy. At high energy, the coupling constant is small, and the quarks are free. On the other hand, at long distance, that is, at low energy, the coupling constant is large and the quarks lose their freeness. With this discovery of the asymptotic freedom, QCD is considered to show well also the properties the deep inelastic scatterings, and the theory of hadrons was established to be the QCD. After that, by calculations on computers, the mass spectra of hadrons are well reproduced in QCD.

On the other hand, in string theory, the dimension of the target spacetime in which strings move should be 26 or 10, otherwise there appears a problem (see Sect. 3.2). Because of this history, the idea using string theory to understand hadrons as an effective theory was away from the mainstream. However, in 1974, J. Scherk and J. Schwarz, and T. Yoneya pointed out that string theory contains gravity interaction, and string theory was reborn as a consistent quantum theory of gravity and as a unified theory. The revolution by D-branes took place in string theory 20 years later, in 1995. The achievement of the D-branes is the point that they can realize non-Abelian gauge theories in 4-dimensional spacetime on their worldvolume at low energy. That is, the foregoing problem of the inapplicability because of the wrong dimensions can be avoided by the advent of the D-branes. And in fact, the point that D-branes are solitons of string theory, that is, at low energy they look like black holes, is a key for accessing the situation with large coupling constants. To look at this, next as a preparation, let us see how non-Abelian gauge theories with large coupling constants admit a stringy description.

---

[12]They were awarded a Nobel prize for their discovery of the asymptotic freedom in QCD.

6.4.2 Large N Limit: From Gluons to String Worldsheet

The gauge symmetry of QCD is SU(3), while if we put N D3-branes parallelly on top of each other then SU(N) gauge symmetry appears. Although QCD has $N = 3$, as a matter of fact, there is a method of evaluation which have a better perturbation theory in this non-Abelian gauge theory as one has a larger N. In standard perturbation theory, scattering processes of particles are described in a power expansion of a small coupling constant g_{YM}, but instead of it, we described it as a power expansion of $1/N$, at large N. This is called "large N expansion" or "$1/N$ expansion." It was shown by G. 'tHooft, in 1974, that interestingly enough this large N expansion in non-Abelian gauge theories gives an interpretation of calculations of Feynman graphs of particles to be worldsheets of strings. It is quite intriguing that in this large N expansion the stringy description naturally appears in non-Abelian gauge theories such as QCD. Since the limit to the large N becomes quite important at its relation to D-branes and black holes which will be described later, let us see the details here.

If we add a gauge field of the "center-of-mass part (of D-branes)" which have appeared in Sect. 5.2 into the SU(N) non-Abelian gauge theory, we have in total N^2 gauge fields (namely gluons). Though I have not mentioned earlier, quarks in QCD are of three kinds, corresponding to the SU(3) gauge symmetry. If we upgrade this to SU(N), we have N kinds of quarks hypothetically.[13]

The bound state made by a pair of a quark and an anti-quark is called meson, and this is a kind of hadrons. The confinement problem of our concern is the question of why the mesons cannot be made separate to a quark and an anti-quark. So, let us depict a situation of a meson moving, by a Feynman graph. Here, for simplicity, we consider a Feynman graph in which a pair of a quark and an anti-quark is created from the vacuum and in the end they happen to pair-annihilate (Fig. 6.14, left). In this Feynman graph, gluons are written by wavy lines, and quarks are by solid lines with arrows (the direction of the arrows stands for the quark or the anti-quark). Now let us draw the gluons by double lines with arrows, instead. Each line is assigned with a number $a, b = 1, 2, \cdots, N$. This double line means that N^2 kinds of gluons are divided into $N \times N$ with one line representing N ways while the other line also representing the other N ways, and the kinds of gluons are described by these two indices $\langle a, b \rangle$. This representation is consistent with group theory (though I omit the details), and here, understand that gluons are represented by double lines $\langle a, b \rangle$ with arrows and, at the interaction vertices where the gluons gather, recognize that we have a rule that these lines smoothly connect with keeping the orientation of the arrows.[14] Then this Feynman graph can be rewritten as Fig. 6.14 right.

---

[13]These three do not mean the species of quarks called for example up quark and down quark. Each species of quarks such as up quark has three kinds. These three kinds are called "color." The reason why the theory of quarks and gluons is called QCD, quantum chromodynamics, is the mechanism of this "color" of the SU(3).

[14]The index $\langle a, b \rangle$ of these gluons resembles the labels of the strings connecting various D-branes which appeared in Sect. 5.2, and as we will see later, indeed these are physically the same.

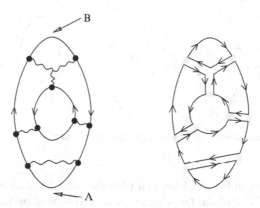

Fig. 6.14 *Left*: A Feynman graph showing a pair of a quark and an anti-quark created in the vacuum and their annihilation back to the vacuum. The *blob* stands for an interaction. At point A, the quark and the anti-quark are pair-created. The vertical direction in the graph is the time direction. At point B they pair-annihilate. *Right*: A redrawn figure in which the wavy lines of the gluons are replaced by double lines with arrows

Let us see how the Feynman graphs with the new writing rules depend on the coupling constant g_{YM} of the non-Abelian gauge theory and the integer N characterizing the gauge symmetry. First, we shall see the dependence of the coupling constant. Among the gluon interaction terms, there are points at which three or four gluons come gather, and it is known that they take values respectively g_{YM} or $(g_{YM})^2$ in the action of the non-Abelian gauge theory. In addition, the value of the interaction at which a gluon comes out from a quark is also g_{YM}. By using these, we find that the g_{YM} dependence of a Feynman graph is

$$(g_{YM})^{V_3 + 2V_4}. \tag{6.20}$$

Here V_3 and V_4 are, respectively, the number of the interacting points appearing in the Feynman graph of the left Fig. 6.14. V_3 counts the number of the cases with three lines gathering at one point, and V_4 counts the case of four lines gathering. Now, suppose that the number of the lines appearing in this Feynman graph (Fig. 6.14 left) is P, then we have a formula $2P = 3V_3 + 4V_4$. This equation can be derived once you notice that each end of the lines are points. With this, the g_{YM} dependence (6.20) is

$$(g_{YM})^{2(P-V)}. \tag{6.21}$$

Here, V is the total number $V_3 + V_4$ of the interacting points.

Next, the N dependence is easier to understood with the double lines with arrows. In this view, N degrees of freedom are given per each line. In the Feynman graph, quarks are pair-created in the vacuum and they pair-annihilate, so any line form a closed loop, and suppose that the number of the loops is I. Then, the N dependence of the Feynman graph apparently becomes N^I. Moreover, if the number of loops

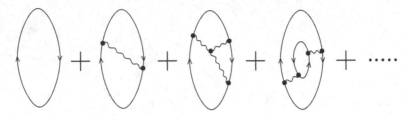

Fig. 6.15 A summation of Feynman graphs in perturbation theory

of quarks appearing in Fig. 6.14 left is h (this includes the outer quark loop), after all, the number $F - 1$ of the faces (the regions surrounded by lines) appearing in Fig. 6.14 left is given $F - 1 = h + I - 1$. (Here, we have defined the number of faces as $F - 1$ for a latter convenience.)

With the previous (6.21), a perturbative summation of Feynman graphs (Fig. 6.15) is given, with coefficients $A_{(P,V,F,h)}$ calculated by the quantum field theory for each Feynman graph, by

$$\sum_{\{P,V,F,h\}} A_{(P,V,F,h)} (g_{\mathrm{YM}})^{2(P-V)} N^{F-h}. \tag{6.22}$$

A statement made by G. 'tHooft is that this formula (6.22) can be seen also as a perturbation theory of string theory. To see this, we reformulate this equation as follows,

$$\sum_{\{P,V,F,h\}} A_{(P,V,F,h)} ((g_{\mathrm{YM}})^2 N)^{(P-V)} \left(\frac{1}{N}\right)^{-2+2g+h}. \tag{6.23}$$

Here we have made a replacement $V - P + F = 2 - 2g$, which you might be familiar with. This is called Euler's polyhedral formula which applies when two-dimensional surfaces are decomposed like a surface of polyhedrons. When any two-dimensional surface is decomposed with arbitrary polygons, with V the number of vertices, P the number of sides, and F the number of faces, then the combination $V - P + F$ does not depend on the way of the polygon decompositions. g is the integer characterizing the continuous two dimensional surfaces in this way, is called a genus. To be more concrete, this g is the number of "holes" of the two-dimensional surface. Spheres have $g = 0$, and the shape of doughnuts (called a torus) has $g = 1$. Note that the vacant space of the torus which the genus counts is different from holes on the surface which can be created by cutting a part of the two-dimensional surface.

Let us see that the summarized (6.23) of the dependence in Feynman graphs coincides with the expansion in the perturbation theory of open string theory. The perturbation expansion of string theory is the expansion creating higher order "Feynman graphs" by adding loops of various open and closed strings to a simple string worldsheet, as in Fig. 6.16. To see the coincidence, let us define the so-called

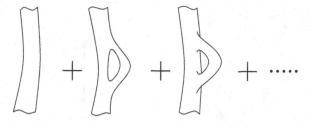

Fig. 6.16 A perturbation expansion in string theory. Attaching closed and open string loops on a worldsheet of a string generates higher order worldsheets in the expansion

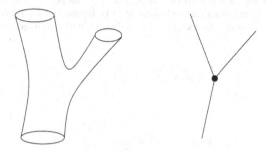

Fig. 6.17 An easier way to see the definition of the string interaction strength g_s is to draw worldlines of particles as in the *right* figure corresponding to the *left* figure of the string worldsheet. The particle view has a single interaction point, correspondingly. In this way, it is a natural idea to assign g_s when a closed string branch comes out

" 'tHooft coupling constant"

$$\lambda = (g_{YM})^2 N. \tag{6.24}$$

We consider a limit which makes N extremely large while keeping λ finite. Then, (6.23) is regarded as a power expansion in terms of $1/N$. Here, if $1/N$ is regarded as the coupling constant g_s of string theory, it just turns out to be a perturbation expansion of worldsheets of strings, as follows.

Let me explain what the string coupling constant is. g_s is, as defined in Fig. 6.17, the extent of branching of string worldsheets, namely, the magnitude of interaction of strings. When a closed string forms a loop, two g_s appear (see Fig. 6.18), so any addition of one loop gives a multiplication of $(g_s)^2$. If the number of closed string loops is g, their effect is $(g_s)^{2g}$. Next, we define the coupling constant of open strings $g_s^{(open)}$ as in Fig. 6.19 left. Then, similarly to the closed strings, a loop of an open string creates a hole, and its effect should be $(g_s^{(open)})^2$. And the coupling constant of closed strings g_s and that of open strings $g_s^{(open)}$ are related via an equation $g_s = (g_s^{(open)})^2$ as is understood in Fig. 6.20. Therefore any open string loop gives a factor g_s appearing. After all, related with the number of hole h, a factor $((g_s^{(open)})^2)^h = (g_s)^h$ should be multiplied. In sum, a worldsheet with genus g and the number of

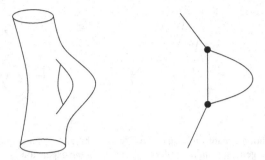

Fig. 6.18 *Left*: A closed string form one loop. This corresponds to creating a single genus. *Right*: You can understand that each worldsheet loop is accompanied with $(g_s)^2$, by looking at a corresponding particle worldlines as it amounts to the case with two interaction points

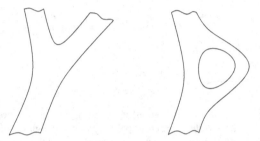

Fig. 6.19 *Left*: the strength of the interaction of open strings, namely, the definition of the open string coupling constant $g_s^{(open)}$. The definition is similar to that of closed strings. *Right*: a loop of an open string creates one hole. $(g_s^{(open)})^2$ is assigned for each hole

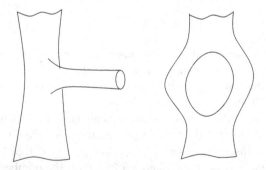

Fig. 6.20 A process of emission of a closed string from an open string (for which g_s is assigned) in the *left* figure can be viewed also as an open string forming a loop as in the *right* figure

holes h has a factor $(g_s)^{2g+h}$. At this stage, if we identify g_s with $1/N$, this is totally equivalent to (6.23). The perturbative expansion by means of the string worldsheets, namely, the expansion by g_s, corresponds to the large N expansion of non-Abelian gauge theories such as QCD.

In this correspondence, we can find that the following rule relates Feynman graphs of QCD to string worldsheets: the lines of quarks are left as they are, and paste worldsheet inside the quark loop lines in the double-line formalism. Namely, the picture is that many gluons fly around and form a worldsheet. For instance, in the case of the Feynman graph of Fig. 6.14, the corresponding string worldsheet is a sphere with two holes, as in Fig. 6.21. This is because in the case of the Fig. 6.14 we have $V_3 = 10, V_4 = 0, V = 10, P = 15, I = 5, g = 0, h = 2$. The graph with all the lines on a plane in the double-line formalism is called a planar graph, whose corresponding worldsheet is a sphere. The ones in Figs. 6.14 and 6.21 are of this kind. In the large N limit, these planar graphs contribute most in the summation of all graphs. And non-planar graphs which do not fit in a plane become worldsheets with nonzero genus g (see Fig. 6.22).

In the correspondence we saw so far, holes stand for worldlines of quarks in Feynman graphs. On the other hand, holes on the worldsheet mean "end points of strings" in the sense that there is a boundary of the worldsheet there. Altogether, we understand that quarks live at the edge of strings. This indeed coincides with the

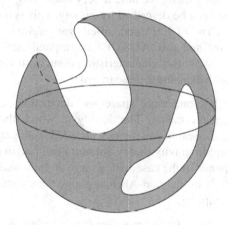

Fig. 6.21 A string worldsheet corresponding to Fig. 6.14. There are two holes on a sphere

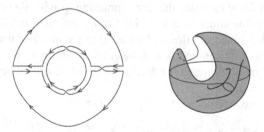

Fig. 6.22 When the double-line graph cannot fit on a plane, the corresponding worldsheet has nonzero genus g. As this graph has $V_3 = 4, V_4 = 0, V = 0, P = 6, I = 1, h = 1, F = 2$, it has $g = 1$

picture introduced in the previous explanation of the confining potential: the gluon interaction stretching straightly between two quarks is interpreted as a string. The string worldsheet picture having appeared in the large N limit of Feynman graphs in QCD gives us a path to regarding string theory as an effective theory of hadrons. Is QCD represented by a string theory, after all? As a matter of fact, the situation with N being large is the case when physics of D-branes can be described by gravity theory. Next, we shall see a surprising duality between gauge theories and gravity theories, brought by D-branes.

6.4.3 Gauge/Gravity Correspondence: Equivalence Between Non-Abelian Gauge Theory and Gravity

In Sect. 5.2, we saw that a non-Abelian gauge theory with a gauge symmetry SU(N) is realized at low energy on N D3-branes put on top of each other. And in Sect. 4.2 we learned that the D3-brane is a classical solution of supergravity theory, that is, a black 3-brane. Combining these, we have a very interesting "conjecture" that non-Abelian gauge theories can be described by gravity. This is a conjecture proposed by J. Maldacena at the end of 1997, and called "gauge/gravity correspondence" or "Maldacena conjecture," and also "ADS/CFT correspondence" by the reason which I will describe later. I show two characteristic points in this conjecture, and from here, let us proceed with explaining these in order.

1. The gravity side has one more spacetime dimensions compared to that of the non-Abelian gauge theory. The duality in which theories with different dimensions are equivalent in this manner is called "holography."
2. The situation where in gravity theory quantum and massive corrections are not necessary corresponds to the case with a large N limit and also a large 'tHooft coupling constant λ, on the non-Abelian gauge theory side. Namely, it is in a region of strong coupling.

Then, using this conjecture, for example we can calculate the potential between a quark and an anti-quark in the non-Abelian gauge theory with a strong coupling, by gravity!

At first, let us imagine what the corresponding gravity theory looks like. As we are considering the situation that N D3-branes are piled up, the corresponding solution is a black 3-brane solution and we need to see the ones with just N electric charges of the Ramond-Ramond field C_{MNPQ} of supergravity theory. The solution has the following metric (though the derivation is not described here)

$$-g_{00} = g_{11} = g_{22} = g_{33} = f(r)^{-1/2},$$

$$g_{44} = g_{55} = g_{66} = g_{77} = g_{88} = g_{99} = f(r)^{1/2}, \qquad (6.25)$$

$$f(r) = 1 + \frac{R^4}{r^4}, \qquad R^4 \equiv 4\pi g_s N(l_s)^4.$$

Here, the D3-branes extend along the direction x^0, x^1, x^2, x^3, and they are at the origin $r = 0$ (r is the radial direction transverse to the D3-branes and is defined as $r \equiv \sqrt{\sum_{i=4}^{9} (x^i)^2}$).[15]

Now, I will show that we need to see the region near $r = 0$, in order to find the correspondence to the non-Abelian gauge theory. We like to know various physical quantities in the non-Abelian gauge theory, and, what is the energy scale of those physical quantities? As we will see, it is in fact related with r. The non-Abelian gauge theory is the degree of freedom of open strings on the D-branes, so, intuitively speaking, we can have an image that if this string has some energy it can move widely and reach far from the D-brane surfaces.

First, to have the non-Abelian gauge theory, we need to take only the massless gauge field among various oscillation modes coming from the open string while throwing away the other massive modes. As you see in the mass formula (3.7), this corresponds to a limit $l_s \to 0$. This is a limit of infinite string tension $\sim 1/(l_s)^2$, and is called a "low energy limit."

With this limit in our mind, we recall the appearance of the non-Abelian gauge theory on D-branes in Sect. 5.2. The string connecting D-branes provide a massless particle when D-branes are on top of each other, but when D-branes are away from each other, the massless particle acquire a mass. From the mass formula (5.5), we find

$$2\pi m (l_s)^2 = \text{distance between D-branes}, \tag{6.26}$$

and this m is the energy scale of the modes of our interest in the non-Abelian gauge theory. Therefore, keeping m finite, we take the above-mentioned low energy limit $l_s \to 0$. We understand that this is a limit where the distance between the D-branes decrease. That is, even though we may make D-branes separate, the distance is extremely small. As the physics of our interest in the non-Abelian gauge theory appears only in the region with this distance, in the black 3-branes solution (6.25) of the corresponding gravity, we have to see the region near $r = 0$. As $r = 0$ is an event horizon of the gravity solution, the limit of looking at the region around there is called a "near-horizon limit."

In the black 3-brane solution (6.25), the r-dependence comes in only as R/r in the function $f(r)$, and the near-horizon limit means that we ignore the part "1+" in the function $f(r)$ while leaving only the second term. Namely, in the near-horizon limit, we may take

$$f(r) = \frac{R^4}{r^4}. \tag{6.27}$$

---

[15]In the solution, the Ramond-Ramond field C_{MNPQ} is nonzero, but it is not written here. The important is that the dilaton field ϕ is constant in this solution. As seen in (4.13), if dilaton is constant, the coupling constant g_s of string theory is constant regardless of the distance from the branes. As you will see next, as we need to see a region close to $r = 0$ in order to find a correspondence to the non-Abelian gauge theory, it helps a lot that g_s is constant and does not diverge to infinity even there.

At this stage, we focus on only the radial direction of the space $x^4, \cdots, x^9$, of the solution (6.25). The angular directions form a five-dimensional spherical surface (written as S^5) surrounding the D3-branes, but it does not appear in the metric (6.25) explicitly. (This way of separation is similar to the situation where, for example in polar coordinates of a two-dimensional plane, separation of the radial and the angular directions provides just a circumference for the angular part.) Omitting writing the angular directions, the metric (6.25) is written as

$$-g_{00} = g_{11} = g_{22} = g_{33} = \frac{r^2}{R^2}, \quad g_{rr} = \frac{R^2}{r^2}. \tag{6.28}$$

This metric is the same spacetime as the metric of the five-dimensional AdS spacetime which appeared in Randall–Sundrum model in Sect. 6.1. This is because, using the coordinate transformation formula (4.5) for the gravity field and regarding $k = 1/R$ after a coordinate transformation $r = Re^{-x^5/R}$, (6.10) is related with (6.28).

In this way, by focusing the region near $r = 0$, the five-dimensional AdS spacetime appears (Fig. 6.23). Although the non-Abelian gauge theory is in a four-dimensional spacetime, the corresponding gravity theory turns out to be in the five-dimensional spacetime including the r direction. The theories with different dimensions are connected, which is exactly the holography (1). In order to define the holography precisely, we need to clarify what physical quantities are evaluated on both sides in what manner. I will describe this later, and before that, we shall see how the condition of the strong coupling in the non-Abelian gauge theory side which was mentioned in (2) appears.

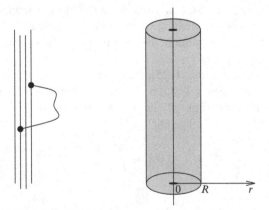

Fig. 6.23 *Left:* A configuration of D3-branes put parallelly. Open strings propagate on those. The vertical direction is x^1, x^2, x^3. *Right:* A conceptual picture of the corresponding black 3-branes. $r = 0$ is an event horizon. The region $r < R$ near the horizon is shaded, which is an AdS spacetime. Outside of it is an almost flat spacetime

6.4.4 Validity of Gravity and Emergence of Large N Limit

Is it safe to look at the region $r < R$ of the black 3-brane solution, first of all? The black 3-brane solution is a solution of the supergravity theory. However this supergravity theory has been derived under the following conditions:

(i) The low energy approximation, namely where we consider $l_s \to 0$.
(ii) We ignore effects of quantum gravity (contribution of Feynman graphs with graviton loops).

Therefore, if these conditions were violated by the classical solution, we cannot trust it, to begin with. So, let us consider in what kind of situation these conditions are satisfied, for the case of the present black 3-brane solution. As for (i), in the solution (6.25) the only parameter which has dimensions is R, so this determines how the gravity field giving the solution is curved. Once the extent of the curving is in a shorter scale than the string length l_s, the solution cannot be trusted. So we need to require $R \gg l_s$, and from the definition (6.25) of R, the condition becomes

$$g_s N \gg 1. \tag{6.29}$$

Here, let us use the relation $g_s = (g_{YM})^2$ between the coupling constant g_{YM} of the non-Abelian gauge theory and the coupling constant of string theory. As I mentioned earlier around Fig. 6.20, first there is a relation $g_s = (g_s^{(open)})^2$, and, the fact that the gluons come out from the open strings deduces that the magnitude of interaction by gluon g_{YM} is equal to the magnitude $g_s^{(open)}$ of the interaction of open strings. From this, the condition (6.29) of (i) is

$$(g_{YM})^2 N = \lambda \gg 1. \tag{6.30}$$

Namely, we have a condition that the 'tHooft coupling constant λ should be large enough.

Next, let us consider (ii). The quantum effects of gravity is given by the coupling constant of gravity, that is, the gravitational constant. The particular length which can be obtained by rewriting the gravitational constant in the dimension of length is caller Planck length, and is written as l_p. This is expressed by the parameters l_s and g_s of string theory, whose derivation is through the derivation of the ten-dimensional supergravity theory at low energy from string theory (I will not describe the derivation here): $(l_p)^8 \sim (g_s)^2 (l_s)^8$. Using this, we find that, for the black 3-brane solution not to be affected by the quantum gravity corrections, we need $R \gg l_p$, namely,

$$N \gg 1. \tag{6.31}$$

This demands the limit of the large N expansion.

Therefore, when we examine the interesting region $r \sim 0$ of the black 3-branes solution, in order for the solution not to violate the validity condition of the derivation of the supergravity theory, we need to require that both the 'tHooft coupling constant λ and N have to be large. It is very interesting that the large N limit naturally appeared. In spite of the discussion of black branes which is totally different from the original discussion by 'tHooft, the same limit appeared.

Is it unlucky that we have the strong coupling condition $\lambda \gg 1$? No, it is not. The gauge/gravity correspondence says that "one can calculate strong coupling region of non-Abelian gauge theories by classical methods in gravity." In Sect. 2.4, I mentioned that the difficulty in calculations in QCD is due to the strong coupling constant. However, if we use the gauge/gravity correspondence, it turns out that we can calculate even the most difficult region of the non-Abelian gauge theory (difficult as seen from the perturbation perspective) by a gravity theory via classical calculations. Then, let us calculate the potential between a quark and an anti-quark which is our concern, by using this correspondence principle.

6.4.5 Deriving Inter-quark Potential by Gravity

Along the 'tHooft's idea, the quark lines in Feynman graphs are holes on worldsheet. Namely, this is the configuration that quarks are put at the end points of an open string. This is natural even in the stand of D-branes. In Sect. 5.2, we considered a brane configuration of Fig. 5.5, where the end points of the D1-brane stand for monopoles. So if we take an S-duality on it, the end points of a fundamental string turn out to stand for electric charges. In QCD, since carriers of the electric charges concerning the gauge symmetry are quarks, we can understand that quarks are put at the end points of the fundamental strings. Furthermore, along the 'tHooft-way of thinking, as there are N kinds of quarks (in the case of QCD, this is SU(3) and so we have $N = 3$, so the quark has three "colors"), there should be N kinds of holes on worldsheets. On the other hand, in the way of thinking of corresponding D-branes, N sheets of D-branes are prepared, and the label $a = 1, 2, 3, \cdots, N$ of the D-branes found in Sect. 5.2 specifies the location of the end points of strings. In this sense too, the picture that quarks are put at the end points of string is consistent.

However, in the current situation of the N D3-branes, there is no fundamental string which has only one end point, so we cannot consider a situation with "a single quark." So, let us consider the situation in Fig. 6.24 left, where one end of a fundamental string is on the N D3-branes while the other is infinitely away from them. This infinitely long fundamental string stands for a single quark.

To evaluate the potential between a quark and an anti-quark, we have to introduce also a single anti-quark. So, let us introduce another infinitely long fundamental string with opposite orientation (Fig. 6.24 left). The opposite orientation means an electric charge of a different sign, as in the case of a D-brane and an anti-D-branes (see Sect. 5.2). And we let the distance between this fundamental string and the anti-fundamental string to be L. The potential between the quark and the anti-quark

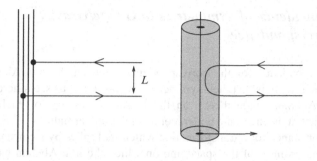

Fig. 6.24 *Left*: A configuration of an infinitely long fundamental string and an anti-fundamental string (with the opposite orientation) are put on the D3-branes at the distance L. In the worldvolume theory of the D3-branes, these strings respectively represent a quark and an anti-quark. *Right*: a corresponding figure of the gravity side. In the shaded region, there exists the gravity field of the AdS spacetime, so the fundamental string can have shorter length and thus a lower energy if it is connected with the anti-fundamental string

should correspond to the energy of the string (the energy of the string simply given by multiplying the length of string by the tension).

We have learned that, in the situation where λ and N is large, the calculation of this configuration in the gravity side by using the gauge/gravity correspondence is a good approximation. So, let us see the appearance of this fundamental and anti-fundamental strings in the gravity side. In the gravity side, the spacetime is curved as given in (6.28). Although the fundamental and the anti-fundamental strings stretch in the direction r, it turns out that they can have lower energy once the strings are connected in the AdS spacetime, actually because of effects of the curved spacetime. As the energy is originally infinite because we have infinitely long strings stretched, we may regard the inter-quark potential energy as the energy gained by the string connection. That is, the problem of the inter-quark potential reduces to a geometrical problem of what configuration of the string minimizes its length in the curved spacetime.

The minimal length problem in the curved spacetime is not easily solved, but one can solve it concretely. Here is the result. The inter-quark potential is given as, after the calculation in the gravity theory,

$$V(L) \sim -\frac{\sqrt{\lambda}}{L}. \tag{6.32}$$

Because this is not proportional to L, it is not a confining potential (6.19). In fact, it is expected that the potential is not a confining potential. I will explain that this is because of a symmetry called "conformal symmetry" which the non-Abelian theory attained on the D3-branes has. Since in the actual QCD there is no conformal symmetry like this, we need to figure out some way to apply this gauge/gravity correspondence to QCD. I will tell the scheme and a derivation of the confining potential at the end of this section.

6.4.6 Coincidence of Symmetries in Gauge/Gravity Correspondence

An important evidence for the correspondence between the non-Abelian gauge theory and the gravity is that their symmetry coincides. Let us see it here.

The non-Abelian gauge theory on the D3-branes on top of each other has a property that it is invariant under conformal transformations. The conformal transformation stands for a transformation which multiplies by a constant the four-dimensional coordinate of the spacetime on which the non-Abelian gauge theory lives,

$$x^\mu \to C x^\mu. \tag{6.33}$$

When a field theory is invariant under this transformation, the theory is called a conformal field theory (which is abbreviated as CFT). On a field $\phi(x)$, in general, the conformal transformation is written as

$$\phi(x) \to C^\Delta \phi(cx). \tag{6.34}$$

Δ is a number given for each field, and called a scaling dimension. To make sure whether a field theory is invariant under this conformal transformation, if quantum effects are not taken into account, it is enough to see the invariance of the action of the field theory. In fact if Δ is taken to be the same as the mass dimension of the field (Here, mass dimensions stand for how to count dimensions with a dimension of mass defined to be $+1$ in the unit system adopted in this book, with the velocity of light $c = 1$. For instance, length has a mass dimension $[-1]$, and energy has a mass dimension $[1]$.), and if in the action there is no constant having a dimension such as the mass m, the action turns out to be conformally invariant. This is because the mass dimensions are originally defined such that the dimension of the total action is zero, and the conformal transformation is just a multiplication by constants according to the mass dimensions. If constants with a mass dimension, such as m, appears in the action, then they are fixed constants so not conformally transformed. However, if some of the quantities do not transform according to their mass dimensions like (6.34), the mass dimension of the action does not vanish, so after all, the action is not invariant.

Now, the non-Abelian gauge theory appearing on the D3-branes on top of each other has no constant with a mass dimension. The only constant is the coupling constant of gauge fields, and its mass dimension is zero. That is, this theory is conformally invariant classically. And it is known that even with quantum effects this theory is indeed conformal invariant. That reason is a huge supersymmetry which this gauge theory has. I mentioned that D-branes correspond to BPS black p-branes in Sect. 4.2, where "BPS" means having unbroken supersymmetries. As a result of that, the non-Abelian gauge theory appearing on the D3-branes also has the large supersymmetries. This non-Abelian gauge theory is called "$\mathcal{N} = 4$ supersymmetric Yang–Mills theory." "Yang–Mills" theory stands for a theory of the gauge field part in non-Abelian gauge theories, and gives us basis of non-Abelian gauge theories. Although $\mathcal{N}$ expresses the number of super-symmetry, the important point is that conformal symmetry is guaranteed in this theory thanks for many super symmetry. The $\mathcal{N} = 4$ supersymmetric Yang–Mills theory is a CFT. From this, the gauge/gravity correspondence is also called "AdS/CFT correspondence."

As the conformal symmetry exists on the non-Abelian gauge theory side, it should be seen also on the gravity side. On the gravity side, the theory is defined by the AdS spacetime (6.28). The conformal symmetry of the gauge theory is understood as a symmetry which does not change the metric of this AdS spacetime. for simplicity, we consider (6.10) which is equivalent to (6.28). Since metric is transformed as (4.5) in coordinate transformations, the conformal transformation (6.33) generates a transformation $g_{11} \rightarrow (1/C^2)g_{11}$. Here, at the same time if we make a transformation

$$x^5 \rightarrow x^5 + (1/k) \log C, \qquad (6.35)$$

the factor $1/C^2$ can be absorbed. That is, the metric (6.10) of the AdS spacetime is invariant if we perform (6.33) and (6.35) simultaneously. Therefore, the conformal symmetry can be seen as a symmetry of the metric in AdS spacetime.

In addition to this, there are other symmetries which coincide on the both sides. For example, in the $\mathcal{N} = 4$ supersymmetric Yang–Mills theory, there are six kinds of scalar fields (and for each we have N^2 species.) You might understand this number six, if you remember that the scalar fields represent the location of the D3-branes in the 10-dimensional spacetime. We choose arbitrary two fields among these six scalar fields, and combine them to form a complex field, and then we make a constant phase transformation. Then it is known that the theory is invariant against this transformation. On the other hand, on the gravity side, remember that there is a five-dimensional sphere in addition to the AdS spacetime. As this five-dimensional sphere does not change under a rotation around an axis of the sphere, so the metric standing for a the sphere does not change. This rotation makes the metric invariant. As you can see here, on the both sides of the gauge/gravity correspondence, the symmetry coincides.[16]

6.4.7 Toward Further Understanding of Quarks

Let us consider why the potential between the quark and the anti-quarks does not become a confining potential. Since this non-Abelian gauge theory has a conformal

---

[16]For readers familiar with group theory, here I will precisely write what the symmetry is. Combining the conformal symmetry and Lorentz symmetry of the $3 + 1$ dimensional spacetime, it turns out to be a symmetry called SO(2,4). This is called a conformal group. SO(2,4) corresponds to a symmetry which does not change the metric of the five-dimensional AdS spacetime (called an isometry). And the transformation of the phase of the scalar fields described here is understood as a group called SO(6) acting on the six scalar fields. This turns out to be an isometry of the metric of the five-dimensional sphere on the gravity side.

symmetry quantum-theoretically, there is no constant with a mass dimension in the theory. When the quark and the anti-quark are put by the distance L in this theory, L turns out to be the unique parameter with a mass dimension $[-1]$. Then, all quantities with mass dimensions have to be written by this constant L, to fit the mass dimensions. On the other hand, the mass dimension of the potential energy V is definitely 1. Therefore, the potential V has to be in inverse proportion to L. That is, the confinement does not occur.

This is a result of the conformal symmetry, and it can be said that (6.32) satisfies it in a right way. Here the interesting is the coefficient of (6.32), that is, a numerator $\sqrt{\lambda} = g_{YM}\sqrt{N}$. Any perturbative calculation of Feynman graphs on the gauge theory side should provide only even powers of g_{YM}, however, (6.32) does not have it. It is considered that, when the coupling constant of the gauge theory is strong, some effects (called non-perturbative effects) which cannot be calculated by the perturbation theory appears and they change the coupling constant dependence. To say more concretely, the coefficient of the potential should be proportional to $(g_{YM})^2$ in the perturbation theory (which is the electrical charge of the quark multiplied by that of the anti-quark), but at the strong coupling, it turns out to be proportional to (g_{YM}) as in (6.32). Since g_{YM} is much less than $(g_{YM})^2$, it is interpreted that the electric charges are screened by the non-perturbative effects at the strong coupling. In this way, by the gauge/gravity correspondence, we can calculate classically the effects which cannot be calculated in the perturbation theory.

The reason why the confining potential does not appear is the conformal symmetry. So, if a confining potential appears in the gauge/gravity correspondence, then we have to adopt what is not an AdS spacetime on the gravity side. It means that we may consider, on the gravity side, a spacetime which does not have a symmetry corresponding to the conformal symmetry. However, you cannot bring in whatever you like. If you do so, we miss what the corresponding gauge theory is. Therefore, various deformations of the D3-branes which we considered have been tried in the gauge/gravity correspondence. Though I cannot describe all the challenge because they are too many, let us finish this section by introducing an important example briefly (Fig. 6.25).

In the foregoing gauge/gravity correspondence, the x^3 direction is along the D3-branes, and this is an infinite length. Here, let us compactify this into a circumference. With this only, symmetries does not break much, so in order to break the supersymmetry, we impose the following boundary condition: when fields go around this circumference once, bosonic fields come back to its original form while fermionic fields acquire a minus sign. This boundary condition breaks completely the supersymmetries which interchanges the bosons and fermions, so this is appropriate for us. With this compactification with making the radius of circumference to be small enough, this theory becomes a Yang Mills theory having no supersymmetry. However, the spacetime dimension turns out to be 3. In this way of the compactification, also in the gravity side, we have to compactify the x^3 direction too. However, as a matter of fact, just the gravity solution of the compactified AdS spacetime is unstable, and it is known that we reach a solution with a deformed central part $r \sim 0$. At large r it is the same as the original AdS

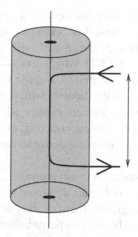

Fig. 6.25 In a deformed AdS spacetime, the connected string has a smaller energy if it sweeps the bottom of the spacetime. Since the length is proportional to the string distance L at the infinity, there appears a confining potential between the quark and the anti-quark

spacetime, but for small r, it deforms and becomes a metric different from the AdS spacetime.

In this spacetime we compute the configuration of a fundamental string such that its energy is minimized, in the same way, then we will find that the confinement occurs. This, to say briefly, is because, since the gravity field around the center is deformed, the string can lose its energy when it goes through a certain value of r (this value depends on your choice of the coordinates, but here let me suppose that it is $r = 0$). The energy is proportional to the length of the strings $r = 0$, which is L. In Fig. 6.24 right, we see that the string hanging down reaches $r = 0$ and there it sweeps at $r = 0$ by the length L. As a result of forming a right angle, the shape of the string turns out to be a part of a rectangle. So the length of the string at $r = 0$ is L. Therefore, as the energy of the string is proportional to L, the confinement potential can be reproduced.

The confinement potential like this has been examined by computer simulations of non-Abelian gauge theories at strong coupling, and is known that there is no contradiction with results using the gauge/gravity correspondence. Calculations for checking the gauge/gravity correspondence have been also made for other various physical quantities.[17] Gravity theories corresponding to gauge theories which are close to QCD have been studied, and recently the mass spectrum of mesons appearing in QCD can be nearly reproduced by the calculations in the gravity

---

[17]For example, in non-Abelian gauge theories, it is known that there are states called glueballs which are bound states of two or more gluons. The mass spectrum of these glueballs calculated by using the gravity theory coincides almost with results of numerical calculations of the spectrum on computers.

theories. However, the evaluation of the spectrum contains various assumptions, and so there are also some issues. For example, as an essential problem, in this gauge/gravity duality it is considered that only the large N limit can make it precise, while in the case of QCD which has $N = 3$, it hardly becomes an accurate correspondence. In spite of that, it is quite interesting and a new physics that classical computations in gravity theories with completely different properties and dimensions can give various hadron masses which cannot be calculated in perturbation theory because of the strong coupling, even with $N = 3 \neq \infty$. There is no doubt that it will provide a new perspective to the elementary particle physics in the future.

The gauge/gravity correspondence is not a proved correspondence. Although in this section we considered the D3-branes and from there we have derived the non-Abelian gauge theories and the gravity theories, first of all, the origin, string theory, has not been exactly formulated beyond its perturbation theory yet. In string theory, methods on how to calculate scattering amplitudes for given states of strings are known in the case of a small coupling constant g_s, which is a perturbation theory. On the other hand, in field theories, solitons are a typical non-perturbative effect. We cannot find solitons forever if we only look at the perturbation theory. Now that D-branes turn out to be the solitons, the definition of string theory beyond its perturbation theory, which can describe strings and D-branes unifiedly, has not been given yet. In the case of field theories, once an action of a field is given, then it gives a perturbation theory and also non-perturbative effects. However, in string theory, a conclusive action corresponding to the field theory action has not been known yet.[18] In the next last chapter, I will finish this book by introducing discussions by present researchers for the big problem on what is the ultimate theory defining string theory.

From Sects. 6.1 to 6.4, various new paradigms of elementary particles physics and cosmology have been described. These, in each subject, often give us not only new visions to re-interpret known physical phenomena, but also an essential understanding and methods which are actually applicable. Moreover, these subjects keep developing with interacting closely with each other. I hope you have grasped that, in elementary particles physics and cosmology, D-branes, the multi-dimensional solitons appearing in string theory, are driving us into a big excitement among researches!

---

[18]The "string field theory" which was mentioned in Sect. 4.1 might become the "field theory action" supposed here. However, in the string field theories, there are various problems: for example, it is difficult to include supersymmetries, and to perform a procedure called "second quantization" which is necessary for any field theory to have a particle picture.

Chapter 7
Toward a Description of Ultimate Theory

In Chap. 6, we saw that drastic changes in various physics around string theory have been brought from string theory, by the discovery of D-branes. Although each of these are very meaningful and interesting developments, on the other hand how about developments in string theory itself? How has string theory been developed by advent of D-branes? String theory is a candidate theory unifying all the elementary particles and their interactions. What extent has string theory been understood as the ultimate theory to?

In field theories, solitons have an important meaning theoretically. As the duality exchanging solitons for elementary particles appearing in various theories is a symmetry exchanging a weak coupling for a strong coupling, it turns out that it is very useful to analyze the cases with strong couplings in field theories, and this becomes a clue to know the structure of the field theories. And, in string theory, D-branes are the solitons (see Sects. 4.1 and 4.2). As a matter of fact, in string theory, as a conjecture derived from supergravity theory, there is a strong-weak coupling duality called S-duality, and it is a symmetry exchanging fundamental strings for D-branes. Then, if we consider D-branes as fundamental constituent elements, can we make a new theoretical formalism with which we can use string theory even in the region of strong couplings of it, that is, a non-perturbative region in which we cannot use the perturbation theory? And does it become the ultimate theory?

String theory is being developed toward possible answers for these questions, currently. The role of D-branes is important and adventurous, to form the basis of string theory as an ultimate theory and even replace the role played by strings. Let us see this development in this chapter. At first, I shall explain a problem that the definitions of string theory is not known yet, and a problem that a principle for determining the vacuum of string theory is missing. It is directly connected with the problem of the spacetime dimension of our world and the problem that the mechanism of the compactification is still missing. Therefore let us see what the spacetime dimension is, from the viewpoint of D-branes. In fact, it will be revealed that the spacetime dimension is not a definite concept in the viewpoint of string theory and D-branes, but an "emergent concept" from physical theories. This thought brings us to "M-theory" in 11-dimensional spacetime, which will be

K. Hashimoto, *D-Brane*, DOI 10.1007/978-3-642-23574-0_7,
© Springer-Verlag Berlin Heidelberg 2012

explained later, and its concrete realization "Matrix theory." This Matrix theory is a theory consisting of D-branes as fundamental constituents! Let us follow surprising developments in the string theory mainstream.

7.1 Definition of String Theory?

In Sect. 2.4, we have learned that, to solve a theory thoroughly, it is important to clarify the symmetry exchanging solitons for elementary particles. When we say "a theory is solved," it means the situation that for a given certain field theory, including all quantum effects, one can determine the vacuum of the theory and calculate the mass spectrum of particles at the vacuum, and information of particle scatterings. These are the most important quantities in elementary particles physics.

However, as for this "solving the theory," string theory has not reached yet even the start point. This is because, as mentioned in the last part of the previous chapter, in string theory the thing corresponding to an action in field theories is not known yet. Namely, this is not a problem on calculation, but a fundamental problem. It is a problem of how to define string theory.

String theory is a theory with strong constraints, and the rules of string perturbation theory (that is, calculations of scattering amplitudes at small coupling constants) are already known. Any definition of string theory beyond the perturbation theory, namely, a "non-perturbative definition" should automatically derive the rules. In the standard case of field theories , once a field theory action is given, we can derive methods for calculating Feynman graphs. On the other hand, in the current string theory, we have to think of it backwards. And as we saw in this book, string theory includes supergravity theory at low energy, and its solitons, the black holes are regarded as D-branes. The information of these solitons should be implemented in any definition of string theory. That is, the S-duality which we saw in Sect. 4.2 has to be derived from the definition of string theory. Saying in other way, the S-duality turns out to be an important point for seeking for the definition of string theory.

And, because string theory has this problem, it leads that string theory also has a serious issue that "the vacuum of string theory has not been determined." Although string theory is considered to be a theory unifying all the forces and the elementary particles, so far string theory cannot answer questions concerning concrete properties of particles, for example, a question like why the electron mass is the value observed. In order to reproduce properties of observed elementary particles as precise as possible, various selections of the internal space in compactification have been tried. However, this may not mean that string theory derives the standard model of elementary particles, but means that we look for a compactification in string theory which suits the standard model. If string theory is the ultimate theory, there should be a mechanism which can determine the internal space compactified, in string theory itself.

Unfortunately, the mechanism has not been found yet. For instance, let us consider the flat 10-dimensional spacetime which is not compactified at all. String theory has no problem there, as the perturbation theory can be written consistently. The flat 10-dimensional spacetime is stable perturbatively. Hence, the problem of the compactification mechanism is a non-perturbative problem beyond the perturbation theory, and it needs a definition of string theory beyond the perturbation theory. To sum all, if we consider string theory as a theory giving the complete explanation of the properties of the actual elementary particles, we need a non-perturbative definition of string theory, inevitably. This is called a problem of vacuum selection. To select a vacuum of string theory means to determine a target spacetime in which strings propagate, which is nothing but to determine the compactified internal space.

Recent accurate observation of cosmic microwave background suggests that a non-zero cosmological constant (the vacuum energy) exists in the universe. So, many research has been carried out to reproduce a 4-dimensional spacetime with a non-zero cosmological constant from string theory or supergravity theory with selected compactifications. According to the results, the number of vacua which are consistent is quite huge, even at some specific setup. This is called a "landscape." The landscape is an approach to seek for a low energy cosmological model close to the reality by adding effects of string theory or D-branes to classical solutions of supergravity theory. So, the problem of this landscape originates in the fact that we miss a non-perturbative definition of string theory.

In the following of this chapter, as a conclusion of this book, I would like to introduce interesting understanding of, and approach to, a non-perturbative string theory, which appeared in researches for solving this big problem, after the appearance of D-branes. Among the various approaches, a common keyword is the "dimension." In the following explanation, you will understand that the definition of string theory should be related closely with a new idea about dimensions There, the multi-dimension solitons of string theory, D-branes, play an extremely important role. D-branes might give a definition of string theory and break new ground for the ultimate theory.

7.2   D-Branes and Their Dimensions

To answer the questions why a compactification occurs and why our spacetime is of four dimensions, first we need to consider in string theory how the spacetime dimension is understood. As a matter of fact, in string theory, the spacetime dimension is a concept which depends on the situation we are dealing with. Let us see this interesting fact by following some examples below.

When we make a Kaluza–Klein mechanism of the compactification in field theories, as we saw in Sect. 3.2, infinite kinds of particles called KK particles appear. Their masses are given by (3.12), and in the case that the internal space is a circumference with a radius R, they are written as

$$m = |s|/R, \tag{7.1}$$

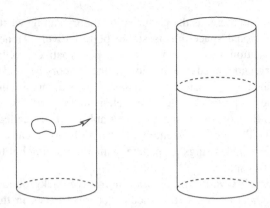

Fig. 7.1 The T-duality. *Left*: a string moves in the compact direction with a momentum in a compactified space. *Right*: a string wrap the compact direction

where s is an integer. Since the KK particles come from a Fourier expansion along the direction compactified, the integer n can be regarded as a momentum of the compactified direction. If we make a KK compactification for massless particles appearing in string theory, in the same way, particle modes with the mass (7.1) appear. Interestingly, in string theory, in addition to them, modes intrinsic to strings appear. They are closed strings winding the compact direction (Fig. 7.1). As the mass of this closed string is the length of the string $2\pi R$ multiplied by the string tension $1/(2\pi(l_s)^2)$, it is, with the winding number w which is an integer,

$$m = |w| R/(l_s)^2. \tag{7.2}$$

In this expression, contributions from oscillations of the string are ignored.

Let us make the following transformation on these (7.1) and (7.2),

$$R \to R' = \frac{(l_s)^2}{R}, \quad s \leftrightarrow w. \tag{7.3}$$

Then, interestingly, the states (7.1) with momenta in the compact direction are exchanged with the states (7.2) with the winding numbers, and the total spectrum turns out to be invariant. Moreover, although an infinite number of modes coming from the oscillations of the string has been ignored here, even with all of them, this invariance still holds. That is, a closed string theory compactified by a radius R is equivalent to a closed string theory compactified by a radius R'. This is called a "T-duality."[1] This transformation (7.3) is a duality transformation because transforming twice brings things back to the original, and "T" is the initial letter standing for a target spacetime.

---

[1]The T-duality was found by K. Kikkawa and M. Yamasaki, in 1984.

What is interesting in this duality is that a compactification with a very small radius R of the circumference gives the same result as that with a large radius compactification. If we take a very small R to try to make the target space dimension decrease, too small radius, as opposed to the expectation, is equivalent to a string theory with a large radius of that direction. That is, only when we look at an energy scale which is lower than both $1/R$ and $1/R'$, we observe a dimension deceasing. However, at an energy scale which is lower than $1/R$ but higher than $1/R'$, the dimension does not appear to decrease, and there appear KK-like particles with masses of order $1/R'$.

What happens for a T-duality in the presence of a D-brane? We will see that, as a matter of fact, the T-duality is a transformation which changes the dimension itself of the worldvolume of the D-brane thoroughly. At first, we consider a situation with a D-brane localized in the compactified space direction (Fig. 7.2 right). Namely, strings satisfy the fixed boundary condition in the direction of the compactification, and strings can wrap the compactified circumference. Let us perform a T-duality transformation to this system. The string with a winding number has to be transformed to a string with a momentum in the compact direction. However, in order to have a momentum in the compact direction for the open string, the direction should take a free boundary condition. That is, the D-brane should extend along the compact direction (Fig. 7.2 left). Therefore, a T-duality along a direction transverse to a D$p$brane changes it to a D$(p + 1)$brane! And as a doubled T-duality transformation brings things back to the original, once in Fig. 7.2 left a T-duality along the D$(p + 1)$-brane is taken, then it becomes the Dp-brane of Fig. 7.2 right. To summarize all, we have the following rules:

- T-duality along D-brane: $Dp \to D(p - 1)$
- T-duality transverse to D-brane: $Dp \to D(p + 1)$

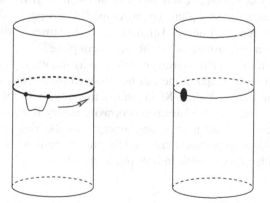

Fig. 7.2 The T-duality on D-branes. *Left*: a D-brane (*thick line*) wrap a compact direction. On the D-brane, an open string move freely in the compact direction, namely, the direction of the D-brane worldvolume. *Right*: a D-brane (*black blob*) is completely localized. A string ending on the D-brane wrap the compact direction

Since this is a duality and so the string theory before and after thetransformation are equivalent, after all, even the dimensions of D-branes change in this way, theories turn out to stand for the same physics.[2]

Let us remember that, in type IIB superstring theory used in this book, Ramond–Ramond fields have even number of indices such as C, C_{MN}, and C_{MNPQ}, therefore only odd numbers for p are allowed for Dp-branes. There, the T-duality transformation brings p to be only even numbers. This theory is called type IIA superstring theory, and at low energy, Ramond–Ramond fields have odd numbers of indices, such as C_M and C_{MNP}. This type IIA superstring theory and the type IIB superstring theory are equivalent to each other, after a compactification at circumferences satisfying a certain relation on the radii (one is R, while the other is R').

In this way, by using the T-duality, D-branes with different dimensions are shown to give equivalent string theories. Then, how about the S-duality? As we saw in Sect. 4.2, in type IIB superstring theory or in type IIB supergravity, the S-duality is a self-duality, and is a duality exchanging fundamental strings for D1-brans in the same theory, as in the duality exchanging solitons for elementary particles. And the S-duality is a duality exchanging weak couplings for strong couplings. Then, in the case of the type IIA superstring theory, what is an equivalent of the S-duality? What is the strong coupling region of the type IIA superstring theory? I will describe next that it actually turns out to correspond to a 11-dimensional theory.

7.3   M-Theory: Emergence of 11-Dimensional Spacetime

In 1995, when a revolutionary paper on D-branes by J. Polthinski was released, another important development was made in string theory. It was an insight by E. Witten that at the strong coupling region of type IIA superstring theory, the space-time dimension increases by one, and then the theory becomes 11-dimensional. Why does the dimension increase by one at the strong coupling?

It is known that from 11-dimensional supergravity theory, type IIA supergravity theory in 10-dimensional spacetime can be obtained by a dimensional reduction, which is a 1-dimensional Kaltza–Klein compactification with the compact radius taken to be zero. The 11-dimensional supergravity theory is known as a beautiful theory with the highest dimension among supergravity theories since before. There appear only two kinds of bosonic fields, and they are a graviton $G_{mn}(x)$ and a gauge field $A_{mnp}(x)$ with antisymmetric indices ($m, n, p = 0, 1, \cdots, 10$). Let us consider

---

[2]Though the discussion here is rather heuristic, actually T-duality can be defined as a transformation exchanging the coordinate τ and σ on the worldsheet of a string, and with it, it can be shown that the free boundary condition and the fixed boundary condition exchange under the T-duality transformation.

the dimension reduction for these fields by following Sect. 3.2,

$$G_{MN} = e^{-(2/3)\phi} g_{MN}, \tag{7.4}$$

$$G_{10\ 10} = e^{(4/3)\phi}, \tag{7.5}$$

$$G_{M\ 10} = -e^{(4/3)\phi} C_M. \tag{7.6}$$

Here, the upper-case indices M, N are for the 10-dimensional spacetime indices running from 0 to 9. With this dimensional reduction, the gravitational part of the 11-dimensional supergravity theory is actually identified with a part of the 10-dimensional type IIA supergravity theory. ϕ is the dilaton field and C_M is the Ramond–Ramond field.

How large is the mass of the KK particles (called "KK gravitons" as they are related to the gravitational parts) appearing in this Kaluza–Klein compactification? The radius of the compactification is determined by the metric of the direction $G_{10\ 10}$, and is given as[3]

$$R \sim \sqrt{G_{10\ 10}} = e^{(2/3)\phi}. \tag{7.7}$$

On the other hand, the mass is measured by the metric (7.4) of the 10-dimensional spacetime. Taking into account the factor of the right side of (7.4), the mass of the KK particle is, with an integer s,

$$m \sim e^{-(1/3)\phi} \frac{|s|}{R} = n e^{-\phi} = \frac{|s|}{g_s}. \tag{7.8}$$

Here in the end we used the relation (4.13) between the dilaton and the string coupling constant. Namely, the mass of the KK particles has g_s in its denominator.

Let us see the meaning of this relations (7.8). First, this KK particle is a solitonic object in string theory. This is because g_s should not appear in the denominator of the mass formula, for string oscillation modes in type IIA superstring theory. Namely, (7.8) is a non-perturbative state, a soliton. The solitons of type IIA superstring theory can be identified with D-branes in the same way for type IIB which we have seen. In the case of type IIA, as p for Dp-branes is even, among them there are D0-branes, which are D-branes having particle-like worldvolumes, and those are identified with the KK particles. In fact, in the process of the dimension reduction (7.6), $G_{M\ 10}$ is identified with a Ramond–Ramond field in ten dimensions, therefore the KK particles with the momentum s of the 11-dimensional direction x^{10} should have an electric charges of this gauge field $G_{M\ 10}$, which is consistent.

The KK particles become lighter as the radius of the compactification gets larger. On the other hand this process corresponds to making the string coupling constant g_s larger, in terms of the D0-branes. Therefore, when the coupling constant grows, the

---

[3]The metric is the scale for measuring the intrinsic length of that direction, and to measure the length of the compactified circumference by using it, we understand the relation between the circumference and the metric.

direction of the 11th dimension emerges! Namely, when the coupling gets stronger in type IIA superstring theory, it is suggested to become an 11-dimensional theory. This 11-dimensional theory is called "M-theory."[4]

The important point in this discussion is that dimensions of spacetime is not an essential element at determining theories, but an emergent concept. When we consider physical theories, first we usually fix the dimension of spacetime and then start discussions. However, in the case of string theory, depending on what kind of situation we consider, the spacetime dimensions themselves change. In the case of type IIA superstring theory, in particular when its coupling constant becomes strong, a new dimension appears. This situation resembles very well that of the gauge/gravity correspondence which we saw in Sect. 6.4. The strong coupling region of non-Abelian gauge theories can be described by gravity theories with spacetime dimensions increased by one. In this way, spacetime dimension is not a definitive concept but an "emergent concept."

Then, what at all is the M-theory? The type IIA supergravity theory is derived by the massless fields of the type IIA superstring theory, then what is the "stringy" theory which derives the 11-dimensional supergravity theory? We don't know the answer to this question yet. However, there is some evidence supporting the idea that this is not a theory of strings stretching in one dimension but of a "membrane" extending in two dimensions.

First of all, in the 11-dimensional supergravity theory, there is a gauge field with three indices, A_{mnp}. Then, applying the discussion of Sect. 4.1, we find that a source (an electric charges) for the gauge field with the three indices is an object having a 2+1 dimensional worldvolume, that is, a "membrane." This membrane is called an "M2-brane." And the object with a magnetic charge turns out to have a worldvolume of 5+1 dimensions in the 11 dimensions, as we understand by using a Hodge duality of Sect. 4.1. This is called an M5-brane. By assuming these two objects, we can show all the players in the type IIA superstring theory appear by the dimension reduction. In the type IIA superstring theory, there are D0-, D2-, D4-, and D6-branes, in addition to fundamental strings and NS5-branes. This NS5-brane is a brane having a magnetic charges of the Kalb–Ramond field B_{MN} (as we have seen in Sect. 4.1, the electric charge is carried by fundamental strings). These various branes are obtained by the dimension reduction from M-theory, as follows:

type IIA superstring theory	M-theory
D0-brane	KK particle
Fundamental string	M2-brane wrapping x^{10}
D2-brane	M2-brane not wrapping x^{10}
D4-brane	M5-brane wrapping x^{10}
NS5-brane	M5-brane not wrapping x^{10}
D6-brane	KK-monopole

---

[4]M-theory was named by E. Witten.

The KK particles have the "electric charge" which is momentum along the direction of the compactification, and the last KK monopole stands for one with its magnetic charge. To say concretely, it is a classical solution of the 11-dimensional supergravity theory and it is smooth at the center.

In this way, assuming M2-branes and M5-branes can let us understand D-branes and fundamental strings unifiedly. This is an extremely attractive idea. However, in the theory of the M2-brane which is a two-dimensional membrane, there is an essential problem that it is quite difficult to quantize oscillation modes of the membrane as opposed to string theory. As quantized oscillation modes have not been obtained, we do not know at all what kind of physics the M2-branes bring us. We do not know even the spectrum of the theory, and cannot calculate scattering processes between the M2-branes. So, we cannot follow the idea of M-theory as a definition of string theory. However, what came from the quantization problem of the membrane theory of the M2-branes was "Matrix theory" which will be described next.

7.4 Matrix Theory: M-Theory Made by D0-Branes

Matrix theory is a theory whose fundamental object is a D0-brane, namely, a KK-graviton. This is a theory of particles, and its concept is as follows. We saw that the 11-dimensional spacetime appears at a strong coupling of string theory. If we want to see this 11-th dimension, it is clear that we should consider various states with momenta of the 11-dimensional direction. So, let us bring many D0-branes and consider a many-body system of them. In particular, supposing that the number of the D0-branes is N, we take a limit $N \to \infty$, then, in the 11-dimensional theory, it should correspond to considering a sector with infinite momentum in the direction of x^{10} (called an infinite momentum frame). Although this does not necessarily mean to see the whole M-theory, we can see at least a part of physics of the 11-th dimension.

Then, what is the physics of the many D0-branes described by? It is the scalar field on the D-branes, as we see the oscillation modes of a string ending on D-branes in Sects. 4.2 and 5.2. Since the D0-branes have 0+1 dimensions for their worldvolumes, this scalar field is function of only time. Moreover there are N^2 kinds of the scalar fields $\Phi(t)$ meaning for the location of the D0-branes, in the present case. (Here we did not count the species coming from the dimensions transverse to the D0-branes.) It is natural to consider a decomposition of this N^2 species into the degree of freedom of $N \times N$, as for the gluons in the 't Hooft large N limit in Sect. 6.4. The scalar fields with indices $\langle a, b \rangle$ $(a, b = 1, 2, \ldots, N)$ can be naturally regarded as a component (a, b) of an N by N matrix. That is, a "matrix-valued scalar field theory in 0+1 dimension" like this is the theory of the many-body system of the D0-branes. This theory is called "M(atrix) theory."[5] The

---

[5] M(atrix) theory was proposed by T. Banks, W. Fischler, S. Shenker, and L. Susskind in 1996, and is called BFSS Matrix theory by taking their initial letters. On the other hand, a matrix model

reason why parentheses are added like "M(atrix)" especially here is that it is for M-theory.

What we have to pay attention to is that the field $\Phi^i(t)$ is a function of only time. That is, this is not a field theory but rather a quantum mechanics standing for motion of point-like particles. The difference from the standard quantum mechanics is that the part corresponding to the location of the particles becomes $N \times N$ matrices. Therefore, a theory for "particles feeling spacetime with matrices" is the M(atrix) theory. It would be exquisitely wonderful if this kind of an interesting generalization of quantum mechanics may turn out to be really proven to represent M-theory.

M(atrix) theory uses the D0-branes (the KK gravitons) as fundamental elements, but M(atrix) theory should be able to describe other objects such as M2-branes. Let us see how the objects with different dimensions, such as M2-branes, are constructed.[6] There, the essence that M(atrix) theory is written by matrices appears.

Let us write the field of the $N \times N$ matrix as $\Phi^i(t)$. Here i is an index standing for the dimensions transverse to the D0-branes. When only the diagonal entries of $\Phi^i(t)$ are non-zero, the meaning of the scalar field (4.12) described in Sect. 4.2 is clear. When the n-th entry among N diagonal ones is written as $\Phi^i_{(n)}(t)$, then it stands for the coordinates of the location of the n-th D0-brane at time t in the spacetime. I will explain why this is the case. The matrix element of the p-th row and the q-th column of the matrix $N \times N$ is the scalar field coming from the oscillation of a string connecting the p-th D0-brane and the q-th D0-brane. Therefore, as the diagonal elements of the matrix correspond to just the case that a string has its ends on the same D0-brane, we can directly use the derivation of (4.12) as we saw in Sect. 4.2, and the value of the diagonal elements stand for the location of the D0-branes.

However, when non-diagonal elements of the matrix are nonzero, this interpretation can not be used. What happens to the D0-branes when the off-diagonal matrix elements appear, showing the essence of what matrices are? As a matter of fact, it turns out to stand for an M2-brane! Then let us see how this solution looks like, concretely. There is a solution of the equation of motion of the M(atrix) theory as follows, (while I will not write the action of the theory,)

$$[\Phi^2, \Phi^3] = 1_{N \times N}. \tag{7.9}$$

Other Φ^i are put to zero. Here, $1_{N \times N}$ in the right hand side is an N by N unit matrix. If we take a trace of the both sides, as we learn easily, the left hand side is zero while the right hand side is N, which is inconsistent, for finite N. However, if N is infinite, it does make sense. Equation (7.9) has the same form as the well-known

---

which provides a definition of type IIB superstring theory was proposed. This is called IKKT matrix model (or type IIB matrix model) which was named after the initial letters of advocators N. Ishibashi, H. Kawai, Y. Kitazawa, and A. Tsuchiya, and researches have been engaged to make clear a mechanism of spacetime generation based on this.

[6]M(atrix) theory is derived by using the method called a matrix regularization, from a theory of an M2-brane (membrane). Here I omit the derivation.

Heisenberg algebra in quantum mechanics,

$$[q, p] = i \frac{h}{2\pi}. \tag{7.10}$$

(q is a coordinate of a particle, p is a momentum of the particle, and h is the Planck constant.) it is known to have a representation (a solution) with matrices of infinite dimensions ($N = \infty$).

For a solution of (7.9), Φ^2 and Φ^3, we can not find one whose nonzero entries are only diagonal ones. This is because it would lead to a vanishing commutator $[\Phi^2, \Phi^3]$. Namely, this is not a situation where D0-branes are localized here and there in the spacetime. Instead of it, infinite number of D0-branes gather and get bound, and they fill the direction x^2 and the direction x^3. This is indeed an M2-brane.

In this way, in M(atrix) theory, in spite of starting with lower-dimensional D-branes, higher-dimensional branes can be constructed. The way to build them is extremely interesting. When many D0-branes come together, not just D0-branes see the spacetime but also string connecting them together see the spacetime. And, by a condensation of the strings (that is, off-diagonal entries of the matrix field Φ become nonzero), higher-dimensional D-branes are formed. From the view point of the D0-branes, spacetime is described by matrix degrees of freedom. This provides us with a completely new perspective of spacetime.

Unfortunately, this very interesting M(atrix) theory has a problem that it is unable to reproduce a situation, for example, where a single fundamental string propagates in the spacetime. However, this example of the M(atrix) theory is quite important in the sense that it gives us a perspective of how D-brane see the spacetime. No matter how M-theory and string theory are formulated non-perturbatively, this way of viewing the spacetime should be included at least as an aspect.

In the M(atrix) theory, D0-branes can construct branes with other dimensions. And, as we saw before, if we use the T-duality, various dimensional D-branes are related with each other, and if we include also the S-duality, even fundamental strings join there. This idea that, whichever D-brane we use as a starting point we finally reach the same complete string theory, is called "brane democracy."[7] As an embodiment of this idea, for example, why can't we construct all by starting from higher-dimensional D-branes? This is a subject which is developing as a discussion like what we saw already in the last of Sect. 6.2. By using unstable D-branes with higher dimensions and letting a tachyon condensation occur, lower-dimensional D-branes are realized as vortex solitons of the tachyon. By preparing infinite pairs of D9-branes and anti-D9-branes which have the highest dimensions for the worldvolumes first, one can create arbitrary number of lower-dimensional D-branes by the tachyon condensation. This idea is called K-theory in mathematics, and it is applied for a classification of possible kinds of D-branes, but we have to wait more development to see it as a non-perturbative string theory.

---

[7]The brane democracy was proposed by P. Townsend in 1995.

7.5 Further Developments in String Theory, for Building Ultimate Theory

In this chapter, my emphasis has been on how the description of spacetime as seen by D-branes and string theory is different from our familiar perspective of spacetime. Spacetime dimension is an emergent notion accompanied by physical theory, and D-branes told us that dimensions vary depending on situations we consider. And, we have learned that D-branes see spacetime in a strange manner via matrices. Physics of D-branes which would challenge our common sense on dimensions, like these, will surely provide us with further surprising physics.

In the 1980s, the ultimate theory unifying all the elementary particles and interactions was considered to be brought from string theory by generalizing point-like particles to strings extending in one dimension. However, at present, we understand that string theory is not a theory just of objects stretching in one dimension, the strings, but a theory controlled by multi-dimensional objects with various dimensions – branes – moving here and there. This theory is not simply that various dimensional objects exist simultaneously, but the whole having every each independent object as a fundamental ingredient which can build others. These objects called (D-)branes have various kinds and different dimensions, and the way they look at the spacetime dimension is extraordinary. In any complete definition of string theory, mutual relations among the whole of these D-branes with various dimensions should be realized clearly. It will be something which unify concepts of different dimensions. Nobody knows what the final definition looks like. However, compared with 20 years ago, it is true that understanding of what the string theory has totally changed. The change means that we gradually come to know the whole view of understanding, including the previous understanding. How does the final definition of string theory contain the recently-found idea of branes and M-theory? How does it relate with the standard model of elementary particles and our observations, as a consistent framework as a whole? To imagine possible answer is very thrilling. And, it may be the case that we just have not recognized it while it is very close to us.

However, it may be opposite and the final definition may not be near us but be very far from us. Some people doubt the meaning of the existence of string theory, and what they may think the biggest question is that, after all, string theory seems not to have sound predictive power. In this viewpoint, one might say that string theory is still far from the goal. We learned in Sect. 3.2 that higher dimensions are required in string theory and in order for it to be consistent with our observable four-dimensional spacetime a part of the spacetime should be compactified tightly. We saw that, as an alternative to the compactification, there is an attractive method called braneworld, in Chap. 6. However, physical mechanism to answer the questions such as why the higher dimensional space is compactified, or why the D-branes exist in the higher dimensional spacetime to give the braneworld, has not been known yet. Therefore, after all, string theory cannot predict what kinds of particles at the low energy region of string theory and what kinds of interaction and mass spectra they should have. This point is, to be said, the biggest mystery of string theory. This is a good question at some extent. As I explained at the beginning of this chapter, this relates with the biggest goal of string theory. The definition of string theory has not been found yet, and the vacuum of string theory is not known.

However it is too early to conclude that, with this fact only, string theory has no meaning for existing. This is because we can say that at present, besides string theory, there is no consistent theory which can calculate quantum corrections of gravity, at all. In this sense, to pursuit string theory is extremely meaningful. However, even with this statement, some people may like to have a concrete meaning via how it can give predictions beyond the standard model, as string theory is there for going beyond the standard model of elementary particles. A gap of "sense" between those who need the ultimate theory and those who need a relation to the present standard model of elementary particles, should be the origin of the question above against string theory. This gap is quite simplistic, and it is important to hold and bind both senses organically, as elementary particle physicists. This gap, so to speak, might resembles the gap which a mathematician feels when he has an interview on winning his Fields Medal: "how is your research useful for our modern society?" Although this example is exaggerated, it is sure that there is some kind of a gap, and it is one of big forces driving researchers of string theory into study.

As achievements of the study, D-branes have provided a new answer concerning the meaning of existence of string theory. It is the fact that D-branes can reveal properties of various kinds of physical theories, in spite of not giving some direct results on the standard model of elementary particles, and the fact that D-branes can give new methods and paradigms for physical theories. This kind of viewpoint is opposite to the view mainly concerning relations to the standard model, so I have not emphasized in this book, However, the stories on the application of D-branes (in Chap. 6) in this book embody this idea quite well. This is the idea that we use string theory and D-branes as a method and as a technology to analyze various physical theories.

For instance, the holography which was mentioned in Sect. 6.4 can be said to be really along this way of thinking. By considering a D-brane configuration corresponding to QCD, and taking a near horizon limit of a classical solutions of gravity, we could calculate various physical quantities which are technically difficult to calculate in QCD. And in Sect. 6.3, in the study of black holes, the entropy is reproduced by bringing D-branes and counting the number of states of strings attached on them. There are also many other examples which have not been mentioned in this book. For instance, physics of solitons and D-branes are very closely related. As we saw briefly in Sect. 5.2, by the D-brane method, in field theories with various spacetime dimensions, new solitons are predicted to exist, and new properties of known solitons are predicted. And they turn out to be proved analytically by field theories analytically or numerically. Although in this book description of a relation between string theory and mathematics of D-branes is omitted, it often happens that by using properties of D-branes, formulas in mathematics and relations among invariants have been predicted, are proved by mathematicians. These are indeed "predictions" made by string theory, and they are successful.

Therefore, we can say that, for the question on the meaning of existence of string theory, D-branes give a little bit exquisite kind of answers. String theory can be directly applied for not only relations to the standard model of elementary particles but also to various other physical theories and mathematics, and it gives predictions, thus is "useful."

Here, on the meaning of research on D-branes, I gave an explanation from a little bit different angle. But of course, the genuine goal of researchers working on string theory is to show that string theory is the theory describing our real world, and to reproduce and include the standard model, and to complete string theory as the ultimate theory in that sense. String theory and D-branes are expanding a certain kind of their "general versatility" which is described above, and developing also their mainstream for approaching a definition of string theory, steadily. I strongly hope that the following framework may work well in the future: the all-purpose versatility in the sense of close relations to various physics described here will produce feedback to string theory from different aspects of physics, with which string theory itself develops. E. Witten, who is a leading scientist of string theory, said in the 1990s that string theory is "a bit of twenty first century physics that somehow dropped into the twentieth century." Now, in the twenty first century, string theory and D-branes take a new step forward, and begin to open new doors to physics.

References

As this book is an introductory book for graduates and undergraduates, I simply give references of frontier researches: published papers. Here is the list of relevant research papers. For each papers, you can find relevant chapters in this book. You will feel that these research papers are really what created new trends in frontier physics.

The numbers after "e-print" mean the preprint numbers at a preprint archive, http://arxiv.org/ from which readers can download all preprints.

1. J. Polchinski, Dirichlet-Branes and Ramond-Ramond charges. Phys.Rev.Lett. **75**, 4724 (1995) e-print: hep-th/9510017. (section 4.2)
2. N. Arkani-Hamed, S. Dimopoulos, G. Dvali, The hierarchy problem and new dimensions at a millimeter. Phys. Lett. **B429**, 263 (1998). e-print: hep-ph/9803315. (section 6)
3. L. Randall, R. Sundrum, A large mass hierarchy from a small extra dimension. Phys. Rev. Lett. **83**, 3370 (1999). e-print: hep-ph/9905221. (section 6.1)
4. G. Dvali, S.-H. Henry Tye, Brane inflation. Phys. Lett. **B450**, 72 (1999). e-print: hep-ph/9812483. (section 6.2)
5. A. Strominger, C. Vafa, Microscopic origin of the Bekenstein-Hawking entropy. Phys. Lett. **B379**, 99 (1996). e-print: hep-th/9601029. (section 6.3)
6. C.G. Callan, J.M. Maldacena, D-brane approach to black hole quantum mechanics. Nucl. Phys. **B472**, 591 (1996). e-print: hep-th/9602043. (section 6.3)
7. J.M. Maldacena, The large N limit of superconformal field theories and supergravity. Adv. Theor. Math. Phys. **2**, 231 (1998). Int. J. Theor. Phys. **38**, 1113 (1999). e-print: hep-th/9711200. (section 6.4)
8. E. Witten, String theory dynamics in various dimensions. Nucl. Phys. **B443**, 85 (1995). e-print: hep-th/9503124. (chapter 7)
9. T. Banks, W. Fischler, S.H. Shenker, L. Susskind, M theory as a matrix model: a conjecture. Phys. Rev. **D55**, 5112 (1997). e-print: hep-th/9610043. (chapter 7)

K. Hashimoto, *D-Brane*, DOI 10.1007/978-3-642-23574-0,
© Springer-Verlag Berlin Heidelberg 2012

Index

AdS/CFT correspondence, 21, 136, 143, 154
AdS spacetime. *See* anti-deSitter spacetime
Anti-D-brane, 94, 115
Anti-deSitter spacetime, 108, 138
Anti-fundamental string, 140
Anti-particle, 65, 94, 121
Asymptotic freedom, 129

Beckenstein–Hawking formula, 120, 122
Big bang, 8, 113, 119
Black brane, 74, 78
  BPS, 86
Black hole, 8, 72, 106, 120
  BPS, 74, 123
  entropy, 120
  evaporation of, 106, 122
  Schwarzschild, 72, 122
  thermodynamics, 120
Boltzmann constant, 120, 122
Boson, 58, 70, 77
Bosonic string theory, 61
BPS, 74, 88
Brane, 1, 74, 103
  — democracy, 157
  — gas cosmology, 119
  — inflation, 114
Braneworld, 7, 20, 49, 100

Classical solution, 22, 73
Co-dimension, 51, 96
Collective motion, 24, 85
Color, 130
Compactification, 63, 147
Conformal field theory, 142
Conformal symmetry, 141

Conformal transformation, 142
Cosmic string, 52, 117
Cosmological constant, 108, 114, 149
Cosmology, 1, 15, 42, 53, 112
  inflationary, 43, 99
Coupling constant, 27, 44, 129
  gauge, 102, 139
  QCD, 46, 126
  string, 81, 133, 139
Cut-off, 102

D-brane, 1, 6, 67, 77
  annihilation of, 62, 93, 95, 116
  bound state of, 86
  collision of, 116, 119
  creation of, 62, 93, 117, 119
Dimensional reduction, 53, 78, 152
Divergence, 5
Duality, 9, 21, 46, 75, 127
  open-closed, 9, 79
  S-, 81, 147
  self, 46, 75, 81
  T-, 150

Einstein–Hilbert action, 71
Electric charge, 44, 69, 102
Electromagnetic field, 62, 68
Electromagnetism, 10, 15, 102
Electron, 3, 22, 44, 68
Electron volt, 101
Elementary excitation, 17, 24, 125
Elementary particle, 2, 15, 24, 27, 100, 148
Elementary particle physics, 2, 23, 41, 100, 148
Entropy, 120, 121

K. Hashimoto, *D-Brane*, DOI 10.1007/978-3-642-23574-0,
© Springer-Verlag Berlin Heidelberg 2012